An Outline of Projective Geometry

An Outline of Projective Geometry

Lynn E. Garner
Brigham Young University

NORTH HOLLAND
New York • Oxford

Elsevier North Holland, Inc.
52 Vanderbilt Avenue, New York, New York 10017

Sole Distributors outside the USA and Canada:

Elsevier Science Publishers B.V.
P.O. Box 211, 1000 AE Amsterdam, The Netherlands

Library of Congress Cataloging in Publication Data

Garner, Lynn E.
An outline of projective geometry.

Bibliography: p.
Includes index.
1. Geometry, Projective. I. Title.
QA471.G25 516.5 80-15152
ISBN 0-444-00423-8

Desk Editor John Haber
Design Edmée Froment
Art Editors Glen Burris and José Garcia
Cover Design Paul Agule Design
Production Manager Joanne Jay
Compositor Computype, Inc.
Printer Haddon Craftsmen

Manufactured in the United States of America

Contents

Preface

Over the years, many excellent textbooks in projective geometry have been published. Many are now "old-fashioned," in that they deal almost exclusively with analytic methods and real geometries; some are primarily synthetic, and overlook analytic methods in favor of "hypermodern" approaches. Many are too broad or specialized to be attractive to university undergraduates.

This text has grown out of a desire to present the fundamentals of projective geometry, using the modern abstract point of view when it offers an advantage and using analytic geometry to illustrate the synthetic ideas. Relationships to elementary geometry, combinatorics, analysis, and linear and abstract algebra have been pointed out where appropriate. Finite geometries and examples from Euclidean geometry are also used in the text.

This book assumes some familiarity with analytic geometry, linear algebra, and introductory modern abstract algebra. A brief treatise on ideas from linear and abstract algebra is presented in the appendix for ready reference.

This text has been developed primarily for junior- and senior-level university students in mathematics and mathematics education, and has proved successful in the classroom. Depending on the instructor's approach, the text will be suitable for undergraduates or beginning graduate students.

An Outline of Projective Geometry

Chapter 1

Incidence Geometry

The subject of geometry is the study of the spatial relationships of objects with one another. In your own studies of geometry heretofore, the ideas of point and line have played a preeminent role. We shall use the concept of incidence structure to combine points and lines into geometric spaces, in which we shall study the properties of points, lines, and other objects relative to one another.

Our main approach is *synthetic*, in that we shall use properties of incidence structures as axioms to develop properties of geometric spaces. We shall also use an *analytic* approach to a large extent, in that we shall use algebraic structures to describe examples of the geometric objects we develop. The synthetic approach will require us to be studiously logical, and often rigorous, as is traditionally the case with geometry. The analytic examples will call on us to combine ideas from analytic geometry, modern algebra, and linear algebra to obtain the desired results.

Section 1.1. Incidence Structures

Definition 1. An *incidence structure* is a triple

$$\sigma = (\mathcal{P}, \mathcal{L}, \mathcal{I})$$

in which $\mathcal{P}$ and $\mathcal{L}$ are sets, $\mathcal{P} \cap \mathcal{L} = \emptyset$, and $\mathcal{I} \subseteq \mathcal{P} \times \mathcal{L}$.

The elements of set $\mathcal{P}$ are called *points*, and the elements of $\mathcal{L}$ are called *lines*. The condition $\mathcal{P} \cap \mathcal{L} = \emptyset$ can then be stated, "No point is a line." We shall use lowercase letters $a, b, c, \ldots, p, q, \ldots, x, y, \ldots$ to denote points, and uppercase letters $A, B, C, \ldots, L, M, \ldots, X, Y, \ldots$ to denote lines. This notation agrees with the usual notation of set theory, for lines in many cases are sets of points. It is also consistent with such writings as those of Dembowski [7] and Stevenson [16].

The set $\mathcal{I}$, since it is a subset of $\mathcal{P} \times \mathcal{L}$, is a relation between $\mathcal{P}$ and $\mathcal{L}$. We call $\mathcal{I}$ the *incidence relation*. If $p \in \mathcal{P}$, $L \in \mathcal{L}$, and $(p,L) \in \mathcal{I}$, we say p *is on* L and also L *is on* p.

Let's consider an example.

Example 1. Let $\sigma = (\mathcal{P}, \mathcal{L}, \mathcal{I})$, with

$$\mathcal{P} = \{a,b,c\},$$
$$\mathcal{L} = \{A,B,C\},$$
$$\mathcal{I} = \{(a,B),(a,C),(b,A),(b,C),(c,A),(c,B)\}.$$

Then σ has three points, a, b, and c, and three lines, A, B, and C. Also, each point is on two of the lines, and each line is on two of the points.

It is customary in geometry to draw figures. We shall follow tradition and draw figures also, but first a word of caution. A figure is a mnemonic device that helps us keep track of the geometrical ideas with which we are dealing. The figure itself is not the object of study, but simply a representation which reminds us what we are studying. Geometry has been called "the art of correct reasoning from incorrect pictures." There is a good deal of truth in that description.

To emphasize the fact that our figures are not themselves the geometric objects we are studying, we belabor a bit the description of the representations we shall use. We represent a point by a dot, such as ·. We represent a line by a long thin mark, such as ———. (Whether lines are straight or not is immaterial; the word "straight" has not yet been defined.) To indicate that a point is on a line, we shall make the dot representing the point incident with the long thin mark representing the line, in this manner: —·—. A dot not incident with a long thin mark, such as ——— ·, represents a point not on the line.

Returning to our example, you can readily see that Figure 1.1 represents the incidence structure described. We can also label the figure to correspond to the incidence structure given, as in Figure 1.2.

Let us introduce some more terminology. If p and q are two (distinct) points of an incidence structure, and there is a line that is on both points p and q, we say p and q *are joined*. If there is exactly one line that is on both p and q, we say p and q *determine a line*; we may denote the line by pq. We also call pq the *join* of p and q.

In Example 1, note for instance that points a and b determine the line C.

If L and M are two lines of an incidence structure, and there is a point that is on both lines L and M, we say L and M *meet*. If there is exactly one point which is on both L and M, we say L and M *determine a point*; we may denote the point $L \cap M$. We also call $L \cap M$ the *intersection* of L and M. Note that here the symbol $L \cap M$ does not necessarily have the same

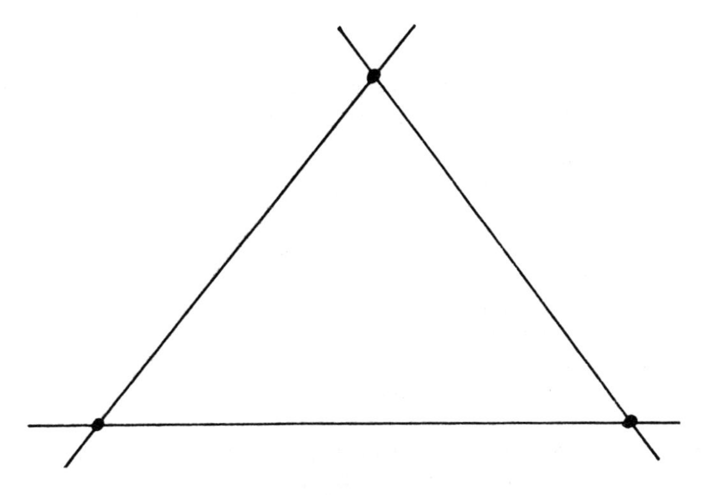

Figure 1.1

meaning as in set theory, for L and M might not be regarded as sets of points.

In Example 1, again, lines B and C meet, and determine point a.

Points that are on the same line are called *collinear* points; the set of all points on line L is called *range* L. Lines that are on the same point are called *concurrent* lines; the set of all lines on point p is called *pencil* p.

Throughout this book, the word "distinct" will be understood whenever a specific number is mentioned. For example, in the phrase "two points a and b," it is understood that $a \neq b$. On the other hand, the phrase "let a and b be points" includes the possibility that $a = b$.

Figure 1.2

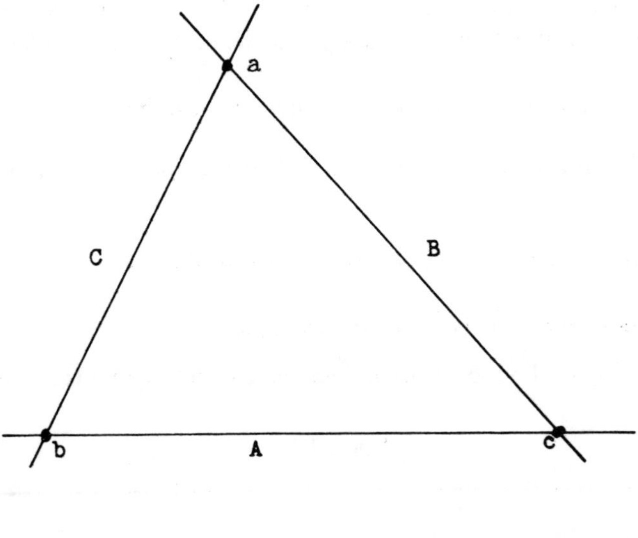

a
C
B
b
A
c

Exercises 1.1

1. In order to become accustomed to the terminology introduced in this section, answer the following questions relative to Example 1.

   (a) Is each pair of points joined?
   (b) Does each pair of points determine a line?
   (c) What is the join of a and c?
   (d) Does each pair of lines determine a point?
   (e) Are points a and b collinear?
   (f) Are a, b and c collinear?
   (g) What is the intersection of B and ab?
   (h) Name the points of the range C.
   (i) Name the lines of the pencil c.

2. Construct and label figures representing all possible incidence structures having exactly two points, p and q, and exactly two lines, L and M.

3. If labels are ignored, are any of the figures in Problem 2 essentially the same?

4. Construct figures representing all the essentially different incidence structures having exactly

   (a) two points and three lines,
   (b) three points and two lines,
   (c) three points and three lines.

5. How many incidence structures, counting even those that are essentially the same, are there that have exactly m points and n lines?

Section 1.2. Planes

Incidence structures form a vast class of geometrical objects, much too broad for us to systematically study all of it. We shall therefore immediately restrict our attention to a class of incidence structures called planes.

Definition 1. A *plane* is an incidence structure satisfying:

Axiom 1. *If p and q are two points, then there is at most one line on both p and q.*

Axiom 2. *If L is a line, then there are at least two points on L.*

To illustrate, consider the following examples.

Example 1. Figure 1.3 represents an incidence structure that is a plane.

Figure 1.3

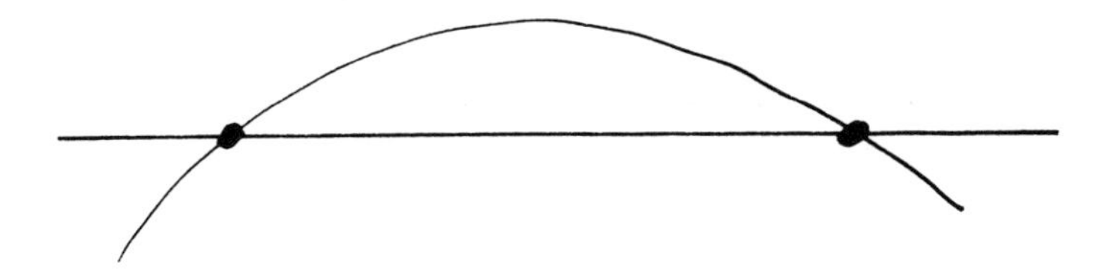

Figure 1.4

Example 2. Figure 1.4 represents an incidence structure that fails to be a plane, because Axiom 1 is not satisfied.

Example 3. Figure 1.5 represents an incidence structure that fails to be a plane, because Axiom 2 is not satisfied.

An immediate consequence of Axiom 1 is stated as our first theorem.

Theorem 1. *If L and M are two lines in a plane, then there is at most one point on both L and M.*

PROOF. Suppose p and q are two points, both on both L and M. Then L and M are two lines on both points p and q, contradicting Axiom 1. Hence there may not be two points on both L and M, and the theorem follows. □

We wish next to define some special types of planes, to which we shall restrict most of our attention. To facilitate their definitions, let us introduce one more term for our technical vocabulary.

If lines L and M in a plane fail to meet, we say L and M are *parallel*.

Definition 2. An *affine plane* is an incidence structure satisfying:

A1. If p and q are two points, then there is exactly one line on both p and q.
A2. If L is a line and p is a point not on L, then there is exactly one line on p that is parallel to L.
A3. If L is a line, then there are at least two points on L.
A4. If L is a line, then there is at least one point not on L.
A5. There is at least one line.

Figure 1.5

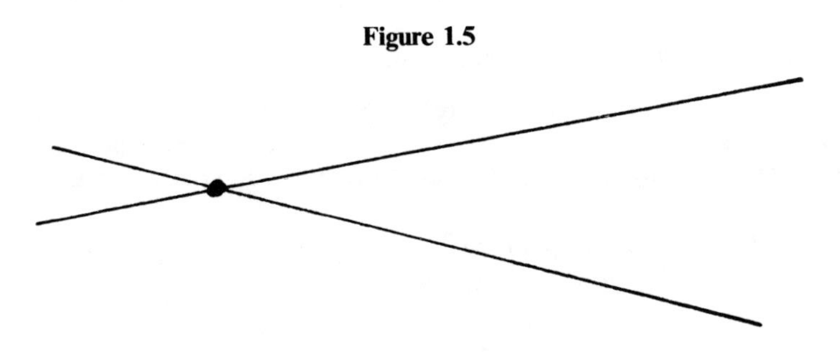

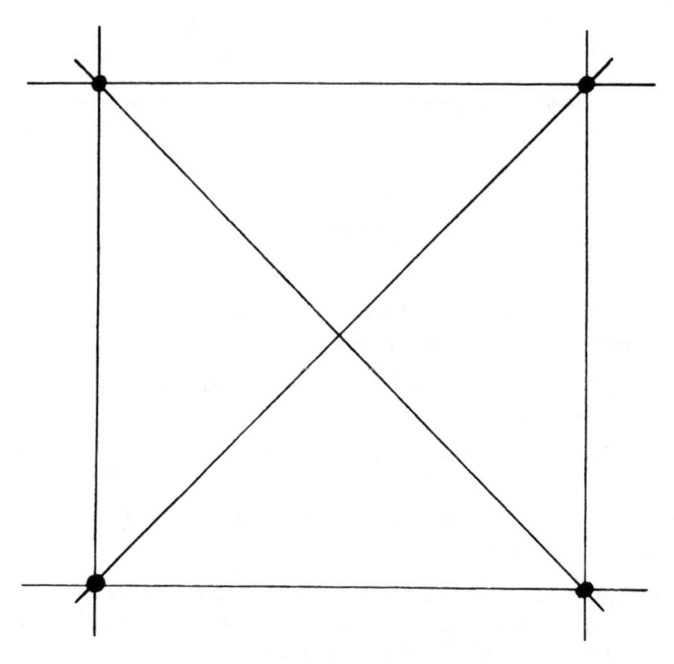

Figure 1.6

It is evident that an affine plane is a plane, for A1 implies Axiom 1 and A3 is Axiom 2. Note also that A1 could be restated, "Two points determine a line."

The axiom A2 of an affine plane is called the parallel postulate. As stated here, it takes the form given by Playfair, which is really the only form of the parallel postulate that is suitable for defining an affine plane. All other forms use notions of distance or angle, neither of which we have yet defined.

Let us look at a simple example of an affine plane.

Example 4. Figure 1.6 represents an incidence structure that is an affine plane. It can readily be shown that this is the "smallest" affine plane which exists.

We now state some simple properties of affine planes as theorems.

Theorem 2. *An affine plane contains a set of three noncollinear points.*

PROOF. By A5, there is a line, say L. By A3, there are two points on L, say p and q, and by A4, there is a point not on L, say r. Then p, q, r are noncollinear, for if they were collinear, the only line on them would be L by A1, and r is not on L. □

Theorem 3. *In an affine plane, if lines L_1 and L_2 are parallel and line $M \neq L_1$ meets L_1, then M meets L_2.*

PROOF. Let M meet L_1 at p. If M is parallel to L_2, then two lines on p are parallel to L_2, contradicting A2. Hence M must meet L_2. □

Another example of an affine plane occurs in the next section.

We now define the type of plane to which we shall devote most of our attention.

Definition 3. A *projective plane* is an incidence structure satisfying:

P1. If p and q are two points, then there is exactly one line on both p and q.

P2. If L and M are two lines, then there is at least one point on both L and M.

P3. If L is a line, then there are at least three points on L.

P4. If L is a line, then there is at least one point not on L.

P5. There is at least one line.

It is evident, as before, that a projective plane is a plane. Notice also that P1 is identical with A1, and can be stated, "Two points determine a line."

Notice that P2 is in marked contrast with A2. We may restate P2, "Two lines always meet." Thus, in a projective plane, there is no such thing as parallel lines. If we combine P2 with Theorem 1, we may conclude the following theorem.

Theorem 4. *In a projective plane, two lines determine a point.*

***Example** 5.* Figure 1.7 represents an incidence structure that is a projective plane. It can be shown that this is the "smallest" projective plane that exists.

An algebraic example of a projective plane occurs in the next section.

We close this section with some theorems stating elementary properties of projective planes.

Theorem 5. *A projective plane contains a set of four points, no three of which are collinear.*

PROOF. By P5, there is a line; call it L. By P3, there are three points on L, say p, a, and b. By P4, there is a point not on L, say c. By P1, there is a line pc, and by $L3$ again, there is a third point d on pc. Now, a, b, c, and d are four points, no three of which are collinear. For, if some three are collinear, then Theorem 4 forces c to be on L. □

Theorem 6. *If p is a point in a projective plane, then there is at least one line that is not on p.*

PROOF. By P5, there is a line, L. If L is not on p, there is nothing to prove. If L is on p, then L has a second point q, by P3, and there is a point r not on L, by P4.

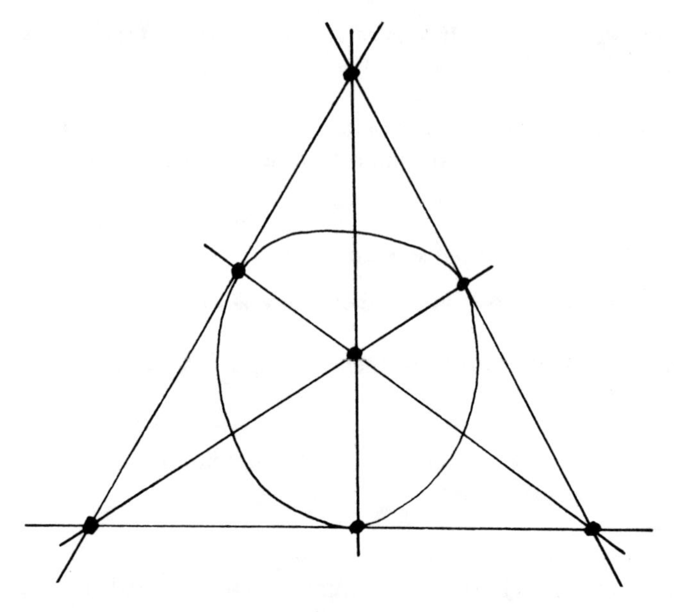

Figure 1.7

Hence by P1 there is a line qr, which is not on p. For if line qr is on p, then Theorem 4 is violated. □

Theorem 7. *If p is a point in a projective plane, then there are at least three lines on p.*

PROOF. By Theorem 6, there is a line L not on p, and by P3 there are three points a, b, c on L. Then by P1, the three lines pa, pb, and pc are on p. That these lines are distinct follows from Theorem 4. □

Exercises 1.2

1. In Exercise 1.1.4, which figures represent planes?

2. Construct a figure representing an affine plane with three points on each line.

***3.** Let α be an incidence structure satisfying the following conditions:

(a) Two points determine a line.
(b) If L is a line and p is a point not on L, then there is exactly one line on p parallel to L.
(c) There is a set of three noncollinear points.

Prove that α is an affine plane.

---

* A starred problem is an important result and is used later in the book.

4. A set of four points no three of which are collinear is called a *four-point*. Prove that every affine plane contains a four-point.

***5.** Let π be an incidence structure satisfying the following:

(a) Two points determine a line.
(b) Two lines always meet.
(c) There is a four-point.

Prove that π is a projective plane.

Section 1.3. Algebraic Examples

This section could rightfully be called "analytic geometry." Our purpose here is to present two basic examples, one of an affine plane, and the other of a projective plane, to which we may refer henceforth. We shall also explore some of the properties of these examples. This section consists of a large amount of algebra and algebraic results, interpreted as geometry.

The first example is the familiar Euclidean plane, presented algebraically. We let R denote the real number system.

Definition 1. The *real affine plane* (also called *Euclidean plane*) is the incidence structure α_R in which

1. each point is an ordered pair (x, y) of real numbers (that is, $\mathcal{P} = R \times R$);
2. each line is the set of points (x, y) satisfying an equation of the form $ax + by + c = 0$, $a, b, c \in R$, and a and b not both equal to zero;
3. point (x_0, y_0) is on line $ax + by + c = 0$ if and only if $ax_0 + by_0 + c = 0$.

To verify that α_R is indeed an affine plane, we employ Exercise 1.2.3. Therefore, we must,

a. given two points (x_1, y_1) and (x_2, y_2), exhibit a line on them, and show that it is unique;
b. given a line $ax + by + c = 0$ and a point (x_0, y_0) not on it, exhibit a line on (x_0, y_0) parallel to $ax + by + c = 0$, and show that it is unique; and
c. exhibit three noncollinear points.

These three things we shall proceed to do.

a. Given points (x_1, y_1) and (x_2, y_2), we note that the line $(y_2 - y_1)x + (x_1 - x_2)y + (x_2y_1 - y_2x_1) = 0$ is on both points. (This line is obtained from the two-point form of the equation of a line.) To show that this line is unique, suppose that $ax + by + c$ is any line on (x_1, y_1) and (x_2, y_2). Then we know that

$$ax_1 + by_1 + c = 0,$$
$$ax_2 + by_2 + c = 0.$$

These two equations can be solved for a and b in terms of c (using substitution, Cramer's rule, or a comparable method); we get

$$a = \frac{c(y_1 - y_2)}{x_1y_2 - x_2y_1}, \qquad b = \frac{c(x_2 - x_1)}{x_1y_2 - x_2y_1}.$$

Hence the line $ax + by + c = 0$ is the line

$$\frac{y_1 - y_2}{x_1y_2 - x_2y_1}cx + \frac{x_2 - x_1}{x_1y_2 - x_2}cy + c = 0,$$

which is equivalent to

$$(y_2 - y_1)x + (x_1 - x_2)y + (x_2y_1 - x_1y_2) = 0,$$

the line we originally displayed.

b. Given the line $ax + by + c = 0$ and point (x_0, y_0) not on it, first note that $ax_0 + by_0 + c \neq 0$, or $c \neq -(ax_0 + by_0)$. Hence the line $ax + by - (ax_0 + by_0) = 0$ is on (x_0, y_0) and parallel to $ax + by + c = 0$. (We may use the idea of slope to see that these lines are parallel, or we may try to solve the two equations simultaneously.) To show that this line is unique, suppose $Ax + By + C = 0$ is any line on (x_0, y_0). Then $Ax_0 + By_0 + C = 0$, or $C = -(Ax_0 + By_0)$. If $Ax + By + C = 0$ is parallel to $ax + by + c = 0$, we must have

$$\begin{vmatrix} a & b \\ A & B \end{vmatrix} = 0$$

(otherwise the two lines meet, by Cramer's rule), or equivalently $aB - Ab = 0$. Since a and b are not both zero, suppose $a \neq 0$. Then $B = (b/a)A$, and the line $Ax + By + C = 0$ is the line $Ax + (b/a)Ay - Ax_0 - (b/a)Ay_0 = 0$, which can be written as $ax + by - (ax_0 + by_0) = 0$, the original line shown.

c. The points $(0,0)$, $(0,1)$, and $(1,0)$ are noncollinear. For if $ax + by + c = 0$ is a line on all three, then we have the conditions

$$c = 0,$$
$$b + c = 0,$$
$$a + c = 0,$$

which imply both a and b are zero, an impossibility.

We have thus verified that α_R is an affine plane. The properties of α_R that we shall use henceforth are the familiar properties of analytic Euclidean geometry; we do not state them separately at this point but simply call them forth as needed.

The second example is not nearly so familiar as α_R, but many of its properties are even more pleasing than those of α_R. As we shall see, this second example is really a generalization or extension of α_R.

Definition 2. The *real projective plane* (also called *extended plane*) is the incidence structure π_R in which

1. each point is a proportionality class $[x_1, x_2, x_3]$ of triples of real numbers, not all zero;
2. each line is a proportionality class $\langle l_1, l_2, l_3 \rangle$ of triples of real numbers, not all zero; and
3. point $[x_1, x_2, x_3]$ is on line $\langle l_1, l_2, l_3 \rangle$ if and only if $l_1x_1 + l_2x_2 + l_3x_3 = 0$.

First some comments on this definition. By a *proportionality class* of triples is meant the set of all triples proportional to a given triple; triple (x_1, x_2, x_3) is *proportional* to triple (y_1, y_2, y_3) in case there is a real number k such that $x_i = ky_i$, $i = 1,2,3$. Thus the point $[x_1, x_2, x_3]$ is the same as the point $[kx_1, kx_2, kx_3]$ for any nonzero k; each point has many representations, or sets of coordinates.

For example, the point $[1, 2, -1]$ can also be represented as $[2, 4, -2]$, or $[-1, -2, 1]$, or $[\frac{1}{2}, 1, -\frac{1}{2}]$, or as $[k, 2k, -k]$ for any nonzero k.

What is true of points is also true of lines in Definition 2. Thus each line has many representations. The zero triple $(0,0,0)$ is not allowed to represent any point or line because of the nature of the incidence relation. The equation $l_1x_1 + l_2x_2 + l_3x_3 = 0$ is called the *incidence condition*. The zero triple would satisfy the incidence condition for every point and for every line, and would create havoc if allowed to represent a point or a line.

To verify that π_R is a projective plane, we shall use Exercise 1.2.5. Then we must

a. given two points $[x_1, x_2, x_3]$ and $[y_1, y_2, y_3]$, exhibit a line on both of them and show that it is unique;
b. given two lines $\langle l_1, l_2, l_3 \rangle$ and $\langle m_1, m_2, m_3 \rangle$, exhibit a point on both of them; and
c. exhibit a four-point.

We shall do these things, at the same time developing some useful results to be used often in working with π_R.

a. Given two points $x = [x_1, x_2, x_3]$ and $y = [y_1, y_2, y_3]$, we first note that the triples (x_1, x_2, x_3) and (y_1, y_2, y_3) are not proportional, so that at least one of the determinants

$$\begin{vmatrix} x_1 & x_2 \\ y_1 & y_2 \end{vmatrix}, \quad \begin{vmatrix} x_1 & x_3 \\ y_1 & y_3 \end{vmatrix}, \quad \begin{vmatrix} x_2 & x_3 \\ y_2 & y_3 \end{vmatrix}$$

is nonzero, say the first. Then

$$\left\langle \begin{vmatrix} x_2 & x_3 \\ y_2 & y_3 \end{vmatrix}, \begin{vmatrix} x_3 & x_1 \\ y_3 & y_1 \end{vmatrix}, \begin{vmatrix} x_1 & x_2 \\ y_1 & y_2 \end{vmatrix} \right\rangle$$

is a line on x and y, for

$$\begin{vmatrix} x_2 & x_3 \\ y_2 & y_3 \end{vmatrix} x_1 + \begin{vmatrix} x_3 & x_1 \\ y_3 & y_1 \end{vmatrix} x_2 + \begin{vmatrix} x_1 & x_2 \\ y_1 & y_2 \end{vmatrix} x_3 = \begin{vmatrix} x_1 & x_2 & x_3 \\ x_1 & x_2 & x_3 \\ y_1 & y_2 & y_3 \end{vmatrix} = 0$$

and

$$\begin{vmatrix} x_2 & x_3 \\ y_2 & y_3 \end{vmatrix} y_1 + \begin{vmatrix} x_3 & x_1 \\ y_3 & y_1 \end{vmatrix} y_2 + \begin{vmatrix} x_1 & x_2 \\ y_1 & y_2 \end{vmatrix} y_3 = \begin{vmatrix} y_1 & y_2 & y_3 \\ x_1 & x_2 & x_3 \\ y_1 & y_2 & y_3 \end{vmatrix} = 0.$$

To show that this line is unique, let $\langle l_1, l_2, l_3 \rangle$ be any line on both x and y. Then

$$l_1 x_1 + l_2 x_2 + l_3 x_3 = 0,$$
$$l_1 y_1 + l_2 y_2 + l_3 y_3 = 0.$$

Since $\begin{vmatrix} x_1 & x_2 \\ y_1 & y_2 \end{vmatrix}$ was assumed to be nonzero, we may solve for l_1 and l_2 in the above system, in terms of l_3;

$$l_1 = \frac{\begin{vmatrix} -l_3 x_3 & x_2 \\ -l_3 y_3 & y_2 \end{vmatrix}}{\begin{vmatrix} x_1 & x_2 \\ y_1 & y_2 \end{vmatrix}} = \frac{\begin{vmatrix} x_2 & x_3 \\ y_2 & y_3 \end{vmatrix}}{\begin{vmatrix} x_1 & x_2 \\ y_1 & y_2 \end{vmatrix}} l_3,$$

$$l_2 = \frac{\begin{vmatrix} x_1 & -l_3 x_3 \\ y_1 & -l_3 y_3 \end{vmatrix}}{\begin{vmatrix} x_1 & x_2 \\ y_1 & y_2 \end{vmatrix}} = \frac{\begin{vmatrix} x_3 & x_1 \\ y_3 & y_1 \end{vmatrix}}{\begin{vmatrix} x_1 & x_2 \\ y_1 & y_2 \end{vmatrix}} l_3.$$

Hence line $\langle l_1, l_2, l_3 \rangle$ becomes

$$\left\langle \frac{\begin{vmatrix} x_2 & x_3 \\ y_2 & y_3 \end{vmatrix}}{\begin{vmatrix} x_1 & x_2 \\ y_1 & y_2 \end{vmatrix}} l_3, \frac{\begin{vmatrix} x_3 & x_1 \\ y_3 & y_1 \end{vmatrix}}{\begin{vmatrix} x_1 & x_2 \\ y_1 & y_2 \end{vmatrix}} l_3, l_3 \right\rangle.$$

Since l_3 is arbitrary (only proportionality counts), we may set

$$l_3 = \begin{vmatrix} x_1 & x_2 \\ y_1 & y_2 \end{vmatrix}$$

and get the line originally displayed.

b. Given lines $L = \langle l_1, l_2, l_3 \rangle$ and $M = \langle m_1, m_2, m_3 \rangle$, it is evident by the same reasoning as in part a that the point

$$\left[\begin{vmatrix} l_2 & l_3 \\ m_2 & m_3 \end{vmatrix}, \begin{vmatrix} l_3 & l_1 \\ m_3 & m_1 \end{vmatrix}, \begin{vmatrix} l_1 & l_2 \\ m_1 & m_2 \end{vmatrix}\right]$$

is on both lines L and M.

c. The points $[1,0,0]$, $[0,1,0]$, $[0,0,1]$, and $[1,1,1]$ constitute a four-point. For if $\langle l_1, l_2, l_3\rangle$ is a line on any three of these points, then $l_1 = l_2 = l_3 = 0$ is the immediate consequence, an impossibility. You are invited to check each of the four cases.

Let us summarize two of the results of the above verification as theorems.

Theorem 1. *In π_R, the line on points $x = [x_1, x_2, x_3]$ and $y = [y_1, y_2, y_3]$ is*

$$xy = \left\langle \begin{vmatrix} x_2 & x_3 \\ y_2 & y_3 \end{vmatrix}, \begin{vmatrix} x_3 & x_1 \\ y_3 & y_1 \end{vmatrix}, \begin{vmatrix} x_1 & x_2 \\ y_1 & y_2 \end{vmatrix} \right\rangle.$$

Theorem 2. *In π_R, the point on lines $L = \langle l_1, l_2, l_3\rangle$ and $M = \langle m_1, m_2, m_3\rangle$ is*

$$L \cap M = \left[\begin{vmatrix} l_2 & l_3 \\ m_2 & m_3 \end{vmatrix}, \begin{vmatrix} l_3 & l_1 \\ m_3 & m_1 \end{vmatrix}, \begin{vmatrix} l_1 & l_2 \\ m_1 & m_2 \end{vmatrix}\right].$$

A particularly convenient way to carry out calculations is afforded by these theorems. To find the line on $[x_1, x_2, x_3]$ and $[y_1, y_2, y_3]$, for example, we may arrange the coordinates of the points in a matrix:

$$\begin{pmatrix} x_1 & x_2 & x_3 \\ y_1 & y_2 & y_3 \end{pmatrix}.$$

Then each coordinate of the line sought is obtained by covering up the corresponding column of the above matrix and reading off the determinant of the remaining columns, changing the sign in the case of the second coordinate.

A simple test for collinearity is contained in the next theorem.

Theorem 3. *Points $[x_1, x_2, x_3]$, $[y_1, y_2, y_3]$, and $[z_1, z_2, z_3]$ in π_R are collinear if and only if*

$$\begin{vmatrix} x_1 & x_2 & x_3 \\ y_1 & y_2 & y_3 \\ z_1 & z_2 & z_3 \end{vmatrix} = 0.$$

The proof of Theorem 3 is left as an exercise, as is the proof of the following useful theorem.

Theorem 4. *Let $a = [a_1, a_2, a_3]$ and $b = [b_1, b_2, b_3]$ be two points in π_R. Point $c = [c_1, c_2, c_3]$ is on line ab if and only if there exist λ, $\mu \in R$, not both zero, such that $c_i = \lambda a_i + \mu b_i$, $i = 1, 2, 3$.*

The pair (λ, μ) of numbers in Theorem 4 is called the pair of *parameters* of c, relative to the *base points* a and b. The assignment to each point on a line L of parameters relative to some fixed base points on L is called a *parametrization* of L. A parametrization of a line is essentially a local coordinate system on the line.

Further properties of π_R will be developed in the exercises.

Exercises 1.3

1. In α_R, name two points on line $ax + by + c = 0$ and one point not on the line.

2. In π_R, name three points on line $\langle l_1, l_2, l_3 \rangle$ and one point not on the line.

3. In π_R, let $a_0 = [1,0,1]$, $a_1 = [0,1,1]$, $a_2 = [1,1,0]$, $a_3 = [1,1,1]$.

(a) Show that a_0, a_1, a_2, a_3 form a four-point.
(b) Find coordinates for the lines a_0a_1, a_0a_2, a_0a_3, a_1a_2, a_1a_3, and a_2a_3.
(c) Find coordinates for the points $d_1 = a_0a_1 \cap a_2a_3$, $d_2 = a_0a_2 \cap a_1a_3$, $d_3 = a_0a_3 \cap a_1a_2$.
(d) Determine whether d_1, d_2, d_3 are collinear.

4. Repeat Exercise 3, with $a_0 = [1,2,0]$, $a_1 = [0,0,1]$, $a_2 = [1,0,-1]$, and $a_3 = [3,-1,-2]$.

5. Prove Theorem 3.

***6.** State and prove a theorem for concurrency of lines in π_R, similar to Theorem 3.

7. In π_R, let $a = [2,-1,1]$ and $b = [0,2,3]$. Find parameters (λ, μ) of c, relative to a and b, for $c =$

(a) $[2,1,4]$, **(b)** $[-2,3,2]$, **(c)** $[2,-1,1]$,
(d) $[0,2,3]$, **(e)** $[4,0,5]$, **(f)** $[6,-1,6]$,
(g) $[8,6,19]$, **(h)** $[10,-13,-7]$, **(i)** $[2\sqrt{2}, -2-\sqrt{2}, -3+\sqrt{2}]$.

8. Prove Theorem 4.

***9.** State a result for lines in π_R, similar to Theorem 4.

***10.** Let a, b, a', and b' be collinear points in π_R, with $a \neq b$ and $a' \neq b'$. Prove that there exists a nonsingular 2×2 matrix M such that if c is any point on line ab, with parameters (λ, μ) relative to a and b, and parameters (λ', μ') relative to a' and b', then

$$(\lambda', \mu') = (\lambda, \mu)M.$$

***11.** Prove that points a, b, and c are collinear if and only if there exist α, β, γ such that $\alpha a_i + \beta b_i + \gamma c_i = 0$.

Section 1.4 Isomorphism

In Exercises 1.1.3 and 1.1.4, we raised the possibility that incident structures can be different, yet essentially the same. That is, they can be alike in structure, but different insofar as labels or interpretations are concerned.

For example, the two diagrams in Figure 1.8 are labeled and oriented differently, but the incidence structures they represent are alike, or are essentially the same. This sameness is made precise in the concept of isomorphism.

Definition 1. Planes $\sigma = (\mathcal{P}, \mathcal{L}, \mathcal{I})$ and $\sigma' = (\mathcal{P}', \mathcal{L}', \mathcal{I}')$ are *isomorphic* if and only if there exist bijections $f: \mathcal{P} \to \mathcal{P}'$ and $F: \mathcal{L} \to \mathcal{L}'$ such that $(p, L) \in \mathcal{I}$ if and only if $(f(p), F(L)) \in \mathcal{I}'$. We call the pair (f, F) of functions an *isomorphism* of σ and σ'. We write $\sigma \sim \sigma'$ to indicate that σ is isomorphic to σ'.

For example, the structures represented in Figure 1.8 can be shown to be isomorphic simply by constructing the isomorphism. We may define f, for instance, by $f: a, b, c, d \to p, r, s, q$, and F by $F: ab, bc, cd, da \to pr, rs, sq, qp$. Then all incidence properties are preserved, as required. (It would not be possible to define $f: a, b, c \to p, q, r$, for b and c determine a line but q and r do not, so no function on lines can preserve the incidence properties.) There are also other isomorphisms possible in this example.

It is clear in the above example that, once f is defined, F is pretty well determined. This is true in general, as the next theorem asserts.

Theorem 1. *If $\sigma = (\mathcal{P}, \mathcal{L}, \mathcal{I})$ and $\sigma' = (\mathcal{P}', \mathcal{L}', \mathcal{I}')$ are planes and $f: \mathcal{P} \to \mathcal{P}'$ is a bijection such that points $a_1, a_2, a_3, \ldots \in \mathcal{P}$ are collinear if and only if their images $f(a_1), f(a_2), f(a_3), \ldots \in \mathcal{P}'$ are collinear, then $\sigma \sim \sigma'$.*

PROOF. We construct a function $F: \mathcal{L} \to \mathcal{L}'$ such that (f, F) is an isomorphism. Let $L \in \mathcal{L}$. Since σ is a plane, there are two points x, y on L. Since x and y are collinear in σ, points $f(x)$ and $f(y)$ are collinear in σ', by hypothesis. Again, since σ' is a plane, there is a unique line L' on $f(x)$ and $f(y)$. Define $F(L) = L'$. Because both σ and σ' are planes, F is well defined.

To show that F is one-to-one, suppose that $F(L_1) = F(L_2)$. If x_1, y_1 are two points on L_1 and x_2, y_2 are two points on L_2, then $f(x_1), f(y_1), f(x_2)$, and $f(y_2)$ are all on the same line. Hence x_1, y_1, x_2, y_2 are collinear, so $L_1 = L_2$. To show that F is onto, let $L' \in \mathcal{L}$. Then L' has two points x', y', and there exist $x, y \in \mathcal{P}$ such that $f(x) = x'$ and $f(y) = y'$, since f is onto. Hence $F(xy) = L'$.

Finally, incidence properties are obviously preserved by (f, F), because of the way F was defined. Hence, (f, F) is an isomorphism, and $\sigma \sim \sigma'$. □

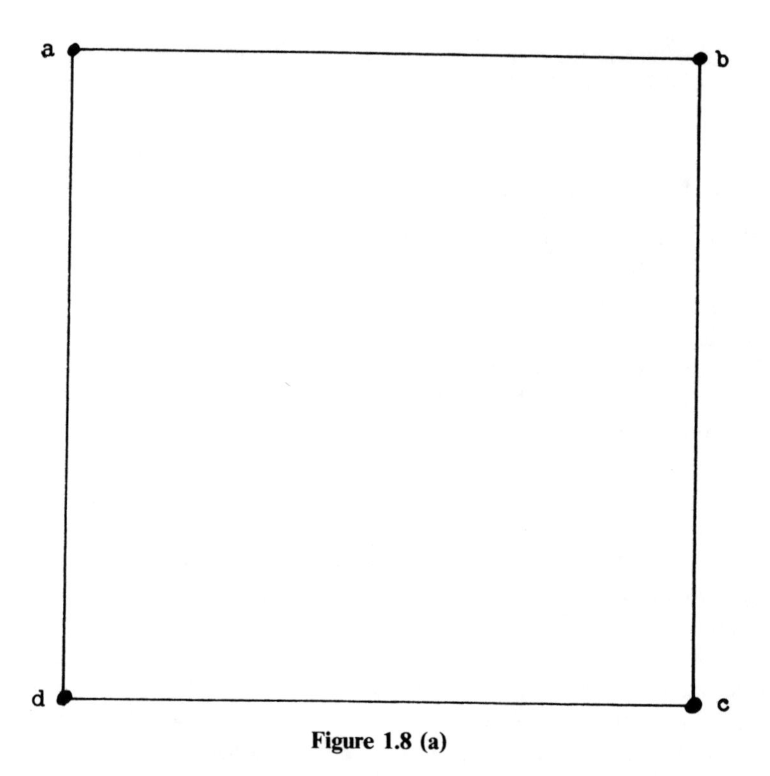

Figure 1.8 (a)

In Section 1.1, we made a distinction between a line and a range, or the set of points on the line. In your experience with Euclidean geometry, however, you ordinarily treated a line as a set of points. Both of these points of view are valid, and in fact are *isomorphic* approaches, as is stated by the following theorem.

Theorem 2. *Any plane is isomorphic to a plane whose lines are sets of points.*

PROOF. Let $\sigma = (\mathcal{P}, \mathcal{L}, \mathcal{I})$ be a plane. For each $L \in \mathcal{L}$, let $L' = \{p \in \mathcal{P} \mid p \text{ is on } L\}$. That is, let L' be the range L. Let $\mathcal{L}' = \{L' \mid L \in \mathcal{L}\}$, and let $\mathcal{I}' = \{(p, L') \mid p \in \mathcal{P}, L' \in \mathcal{L}', \text{ and } p \in L'\}$. Set $\sigma' = (\mathcal{P}, \mathcal{L}', \mathcal{I}')$. Now let $f: \mathcal{P} \to \mathcal{P}$ be the identity mapping. The function f now clearly satisfies the hypotheses of Theorem 1, so $\sigma \sim \sigma'$. □

The plane $\sigma' = (\mathcal{P}, \mathcal{L}', \mathcal{I}')$ might also be denoted $\sigma' = (\mathcal{P}, \mathcal{L}, \in)$, where $\in$ is used to symbolize the incidence relation defined by set membership. Whenever we denote a plane by $\sigma = (\mathcal{P}, \mathcal{L}, \in)$, we are implicitly stating that the lines of σ are sets of points, and "on" is determined by set membership. For instance, the plane α_R is defined in that way.

In some of our work we shall make a distinction between lines and

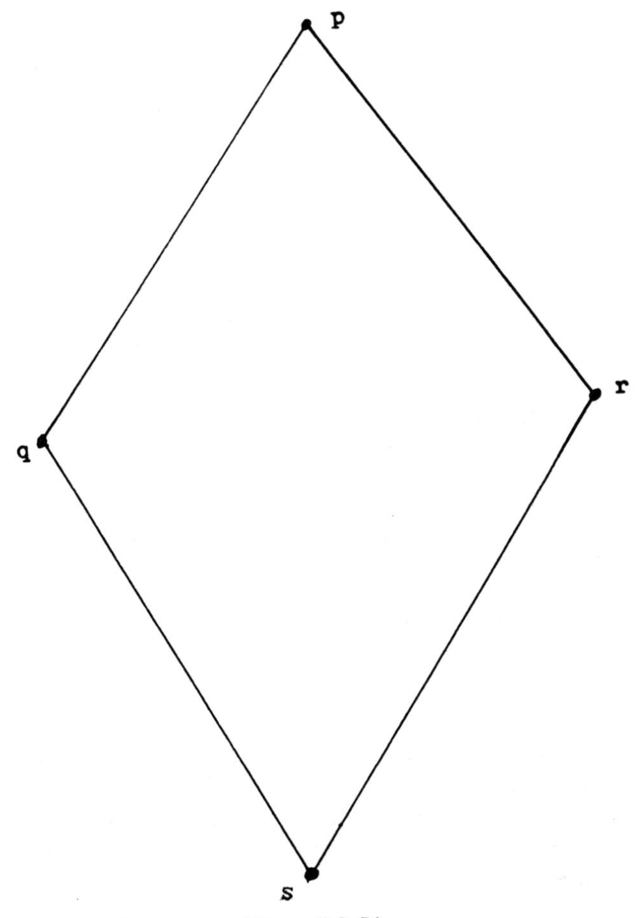

Figure 1.8 (b)

ranges, and in other parts of it we shall not. The result of Theorem 2 is that we may or may not identify lines and ranges, as we choose.

The idea of isomorphism pervades all of modern mathematics. Different structures, whether geometric, algebraic, or analytic, that are isomorphic have all the same properties. Hence, in what follows, we can completely describe all the structures isomorphic to a given one simply by describing one of them.

Exercises 1.4

1. Let S be a sphere in Euclidean space of three dimensions. Let π_S be the plane whose "points" are the diameters of S and whose "lines" are the great circles of

S; we say point d is *on* line C if and only if d is a diameter of C. Show that plane π_S is a projective plane isomorphic to π_R.

2. Let $\pi = (\mathcal{P}, \mathcal{L}, \mathcal{I})$ and $\pi' = (\mathcal{P}', \mathcal{L}', \mathcal{I}')$ be projective planes, and let $f: \mathcal{P} \to \mathcal{P}'$ be a bijection such that if $a_1, a_2, a_3 \in \mathcal{P}$ are collinear, then $f(a_1), f(a_2), f(a_3) \in \mathcal{P}'$ are collinear. Is $\pi \sim \pi'$?

Section 1.5. Duality

In the plane π_R, everything that is true of points seems to be true of lines also. This fact is no accident, but is an effect of a property called duality.

Definition 1. Let $\sigma = (\mathcal{P}, \mathcal{L}, \mathcal{I})$ be an incidence structure. The *dual* of σ is the incidence structure $\sigma^d = (\mathcal{L}, \mathcal{P}, \mathcal{I}^{-1})$. That is, points of σ^d are lines of σ, lines of σ^d are points of σ, and $\mathcal{I}^{-1} = \{(L, p) \mid (p, L) \in \mathcal{I}\}$.

Example 1. Let $\sigma = (\mathcal{P}, \mathcal{L}, \mathcal{I})$, with $\mathcal{P} = \{a, b\}$, $\mathcal{L} = \{L\}$, $\mathcal{I} = \{(a, L), (b, L)\}$. Figure 1.9 represents σ. Then σ^d is represented by Figure 1.10. Note that σ is a plane, but σ^d is not.

Definition 2. Let S be a statement about points and lines in an incidence structure. The *dual* of S is the statement S^d obtained from S by interchanging the terms "point" and "line."

It is evident that if the statement S is true of the incidence structure σ, then its dual S^d is true of σ^d. For instance, in Example 1, the statement "σ has two points" is true, and its dual, "σ^d has two lines," is also true. Similarly, "Each line is on two points" is true of σ, and its dual, "Each point is on two lines," is true of σ^d.

Definition 3. Let C be a class of incidence structures. We say *the principle of duality holds* in C in case the dual of each structure in C is a structure in C.

For example, the principle of duality fails to hold in the class of planes, as indicated in Example 1. An important positive example is stated in the following theorem.

Theorem 1. *The principle of duality holds in the class of projective planes.*

Figure 1.9

a b L

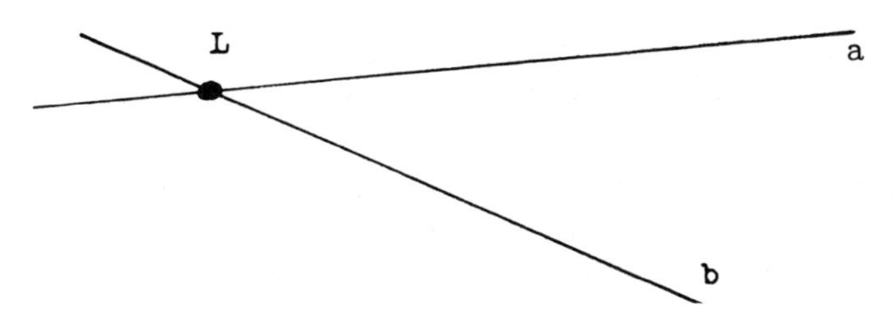

Figure 1.10

PROOF. Let π be a projective plane. We must show that π^d is also a projective plane. By definition, π satisfies the statements P1–P5. Hence, π^d satisfies the following statements:

P1^d. If L and M are two lines, then there is exactly one point on both L and M.
P2^d. If p and q are two points, then there is at least one line on both p and q.
P3^d. If p is a point, then there are at least three lines on p.
P4^d. If p is a point, then there is at least one line not on p.
P5^d. There is at least one point.

The plan of our proof is to show that π^d is a projective plane by proving that if P1^d–P5^d hold, then P1–P5 also hold. We will verify P1–P5 one at a time.

P1. P2^d gives us at least one line on two given points. Suppose L and M are two lines on the two points p and q. Then p and q are on both L and M, violating P1^d. Hence at most one line can be on two given points, and P1 follows.

P2. P1^d implies P2 directly.

P5. P5^d, together with either P3^d or P4^d, implies the existence of a line.

P4. If L is a line, then P5^d implies the existence of a point p. If p is not on L, we are done. If p is on L, then there is a line M not on p, by P4^d. Also, there is a second line N on p, by P3^d. Now there is a point q on both M and N, by P1^d, and q is not on L. For if q is on L, then both L and N are on p and q, contradicting P1, which we have already shown to hold.

P3. If L is a line, then there is a point p not on L, by P4, which we have just proved. By P3^d, there are three lines A, B, C on p, and by P1^d, there are three points $L \cap A, L \cap B, L \cap C$ on L.

Thus, we have shown that π^d also satisfies P1–P5, so π^d is a projective plane. □

ALTERNATE PROOF. Let π be a projective plane. We shall show that P1^d–P5^d hold in π, and hence P1–P5 hold in π^d. [It is obvious that $(S^d)^d = S$.]

P1^d is Theorem 1.2.4.
P2^d is implied by P1.
P3^d is Theorem 1.2.7.
P4^d is Theorem 1.2.6.
P5^d is a consequence of Theorem 1.2.5.

Hence P1–P5 hold in π^d, and π^d is a projective plane. □

The value of having the principle of duality hold in a class of incidence structures is given in the following metatheorem (theorem about theorems).

Theorem 2. *If C is a class of incidence structures in which the principle of duality holds, and S is a theorem which is true of every structure in C, then S^d is also a theorem which is true of every structure in C.*

The proof of this metatheorem is left as an exercise. Because of this result, we automatically have the dual of any theorem we can prove whenever the principle of duality holds.

Because of this, the principle of duality is a sort of "two for the price of one" principle.

Definition 4. A plane σ is *self-dual* in case σ is isomorphic to its dual σ^d.

Theorem 3. *If σ is a self-dual plane and S is a theorem of σ, then S^d is also a theorem of σ.*

PROOF. If $\sigma \sim \sigma^d$ and S is true of σ, then S is also true of σ^d. Hence S^d is true of σ^{dd}. □

Exercises 1.5

1. Let C be the class of incidence structures satisfying Axiom 1 of Definition 1.2.1. Prove that the principle of duality holds in C.

2. Determine whether the principle of duality holds in the class of affine planes.

*3. Let C be the class of projective planes for which a given statement S is true. Suppose that S implies S^d. Does the principle of duality hold in C?

4. Prove that π_R is self-dual.

5. Prove Theorem 2.

6. The terms "point" and "line" are called *dual terms*, because they are interchanged in going to the dual of a statement. "Collinear" and "concurrent" are another pair of dual terms. How many other pairs of dual terms can you find?

7. Can you explain why a distinction is made between "line" and "range," particularly when duality holds?

Section 1.6. Configurations

Definition 1. A *configuration* is a plane $\sigma = (\mathcal{P}, \mathcal{L}, \mathcal{I})$ in which $\mathcal{P} \cup \mathcal{L}$ is finite.

That is, a configuration is a finite plane. We start out by looking at several examples of configurations. These examples will be described by incidence tables, in which the points are listed down the side of the table, lines are listed across the top of the table, and an $\times$ in a certain position indicates that the point to the left of that position is on the line above that position.

Example 1. A *triangle* has the incidence table

	A	B	C
a		$\times$	$\times$
b	$\times$		$\times$
c	$\times$	$\times$	

The points a, b, c are called *vertices* and the lines A, B, C *sides* of the triangle.

Example 2. A *complete four-point* has the incidence table

	L_1	L_1'	L_2	L_2'	L_3	L_3'
a_0	$\times$		$\times$		$\times$	
a_1	$\times$			$\times$		$\times$
a_2		$\times$	$\times$			$\times$
a_3		$\times$		$\times$	$\times$	

The points a_0, a_1, a_2, a_3, called *vertices*, form a four-point. The lines are called *sides*; the pairs L_1 and L_1', L_2 and L_2', L_3 and L_3' are called pairs of *opposite* sides. A point, if any, that is on both of a pair of opposite sides, is called a *diagonal point*.

Example 3. A *Fano configuration* has the incidence table

	L_1	L_1'	L_2	L_2'	L_3	L_3'	D
a_0	$\times$		$\times$		$\times$		
a_1	$\times$			$\times$		$\times$	
a_2		$\times$	$\times$			$\times$	
a_3		$\times$		$\times$	$\times$		
d_1	$\times$	$\times$					$\times$
d_2			$\times$	$\times$			$\times$
d_3					$\times$	$\times$	$\times$

The Fano configuration is recognizable as a complete four-point with three collinear diagonal points.

Example 4. A *Pappus configuration* has the incidence table

	L	L'	A	A'	B	B'	C	C'	P
a	×				×		×		
b	×		×					×	
c	×			×		×			
a'		×				×		×	
b'		×		×			×		
c'		×	×		×				
a''			×	×					×
b''					×	×			×
c''							×	×	×

The points a'', b'', c'' can be constructed from a, b, c and a', b', c', and all lie on P, which is called the *Pappus line* of the two triples a, b, c and a', b', c'.

Example 5. A *Desargues configuration* has the incidence table

	A	B	C	A'	B'	C'	A''	B''	C''	L
a		×	×				×			
b	×		×					×		
c	×	×							×	
a'					×	×	×			
b'				×		×		×		
c'				×	×				×	
a''	×			×						×
b''		×			×					×
c''			×			×				×
o							×	×	×	

The Desargues configuration is recognizable as two triangles abc and $a'b'c'$, with lines aa', bb', and cc' concurrent and points $ab \cap a'b'$, $ac \cap a'c'$, and $bc \cap b'c'$ collinear.

Definition 2. A configuration is said to be *tactical* in case there exist integers r and s such that each point is on exactly r lines and each line is on exactly s points. If a tactical configuration σ has m points, each on r lines, and n lines, each on s points, we say the *form* of σ is (m_r, n_s). We abbreviate (m_r, m_r) to (m_r).

It is easily seen that the foregoing examples are all tactical configurations. The triangle has form (3_2), the complete four-point has form $(4_3, 6_2)$, the Fano configuration has form (7_3), the Pappus configuration has form (9_3), and the Desargues configuration has form (10_3).

The value of the form of a tactical configuration lies in the fact that certain configurations can be characterized by their forms. It is not in general the case that only one plane (up to isomorphism) has a given form. It is true that all configurations with form (7_3) are isomorphic to the Fano configuration, but there are three nonisomorphic configurations with form (9_3), and ten with form (10_3). The following theorems stipulate some very useful characterizations that can be made.

Theorem 1. *If σ is a tactical configuration with form (m_n), with $m = n^2 - n + 1$ and $n \geqslant 3$, then σ is a projective plane.*

PROOF. We shall verify the axioms P1–P5. First, let p be any point of σ. Since p lies on exactly n lines, and since each line is on exactly n points ($n - 1$, not counting p), there are exactly $n(n-1) = n^2 - n$ points that are joined to p, other than p itself. But since there are only $n^2 - n + 1$ points, that means that every point different from p is joined to p. Thus any two points are joined. Since σ is a plane, at most one line is on both of two points. Hence P1 holds.

Next, let L be any line in σ. There are exactly n points on L, and $n - 1$ lines distinct from L on each of those points, so $n(n-1) = n^2 - n$ lines meet L. Hence every line distinct from L meets L, so two lines always meet, and P2 holds.

P3 holds, since $n \geqslant 3$, and P4 holds, since $n^2 - n + 1 > n$ for $n \geqslant 3$. P5 holds because $n^2 - n + 1 \geqslant 1$.

Thus σ is a projective plane. □

Theorem 2. *A finite projective plane is a tactical configuration, and, for some $n \geqslant 3$, has form (m_n), with $m = n^2 - n + 1$.*

PROOF. Let π be a finite projective plane, and let L be a particular line in π. Since π has a finite number of points, line L has a finite number of points on it, say n. Then $n \geqslant 3$ by P3. We shall now prove some claims, based on the fact that line L is on exactly n points.

Claim 1. Every line of π is on exactly n points.

PROOF. Let M be any line of π. If $M = L$, we are done. If $M \neq L$, there is still a point on both L and M; call it p_1 (see Figure 1.11). Let $p_2, \ldots, p_n$ be the remaining points on L. Now M has a second point q, and line $p_n q$ has a third point r. Now each line rp_i meets M, at point q_i, say. Then $q_1 (= p_1), q_2, \ldots, q_n (= q)$ are n distinct points on M, and M is on at least n points. If x is any point on M, then line rx meets L, say at p_i, and $x = q_i$. Hence M has at most n points, so M is on exactly n points.

Claim 2. Every point of π is on exactly n lines.

PROOF. Let p be any point of π, and let M be a line not on p. (M exists by $P4^d$.) Let $q_1, \ldots, q_n$ be the n points on M (Claim 1). Then each of the lines $pq_1, \ldots, pq_n$ is on p, so p is on at least n lines. Moreover, any line on p must meet M, say at q_i, and so is the line pq_i. Hence at most n lines are on p.

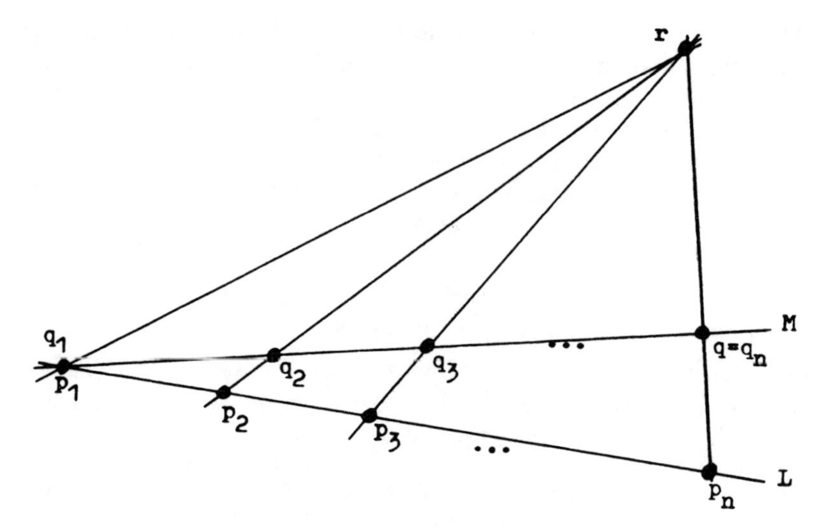

Figure 1.11

Figure 1.12

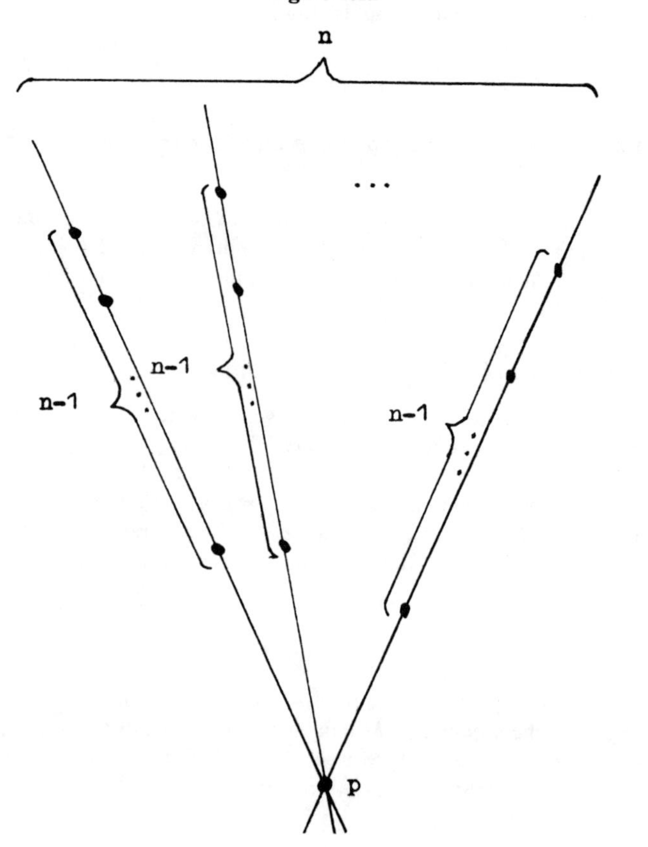

Claim 3. There are exactly $n^2 - n + 1$ points in π.

PROOF. Let p be a point in π. Since every point is joined to p, we may count the points by counting the points on lines through p. Since there are exactly n lines on p, and on each of those lines there are $n - 1$ points distinct from p (see Figure 1.12), there are exactly $n(n-1) + 1 = n^2 - n + 1$ points.

Claim 4. There are exactly $n^2 - n + 1$ lines in π.

PROOF. By duality, from Claims 2 and 3.

By Claims 1 and 2, π is tactical. From all four claims, we see that π has the form $(n^2 - n + 1_n)$. □

Theorem 3. *If σ is a tactical configuration, with form $(n^2_{n+1}, n^2 + n_n)$, then σ is an affine plane.*

Theorem 4. *A finite affine plane is a tactical configuration, and for some $n \geqslant 2$ has form $(n^2_{n+1}, n^2 + n_n)$.*

The proofs of Theorems 3 and 4, which parallel the proofs of Theorems 1 and 2, are left as exercises.

Exercises 1.6

1. Construct figures representing the configurations of Examples 1–5. Wherever possible, use "straight" lines.

2. Construct figures representing the duals of the configurations of Examples 1–5. Again, use "straight" lines whenever possible.

3. Which of the configurations of Examples 1–5 are self-dual?

4. Does the principle of duality hold in the class of tactical configurations?

5. In π_R, let $a = [1,2,0]$, $b = [0,1,1]$, $c = [2,5,1]$, $a' = [1,-1,0]$, $b' = [2,1,1]$, $c' = [1,2,1]$ in a Pappus configuration. Find coordinates for a'', b'', c'', and verify that they are collinear.

6. In π_R, let $a = [1,-1,0]$, $b = [2,1,0]$, $c = [-4,1,2]$, $a' = [2,-2,1]$, $b' = [2,1,-1]$, $c' = [4,-1,1]$ in a Desargues configuration. Verify that aa', bb', cc' are concurrent. Find coordinates for a'', b'', c'', and verify that they are collinear.

7. Construct figures and/or incidence tables describing the tactical configurations having the following forms:

   (a) (n_2), with $n \geqslant 3$; **(b)** $(10_2, 5_4)$,
   (c) $(5_4, 10_2)$, **(d)** $(9_4, 12_3)$,
   (e) (8_3), **(f)** (13_4).

8. Prove that the Fano configuration is a projective plane.

9. A gardener wishes to plant a grove of ten ornamental trees in five rows with four trees in each row. Can he do it?

10. The members of a boys' club decided to form themselves into five-man basketball teams in such a way that any given pair of boys in the club would be teammates on exactly one team. After the team rosters were drawn up, it was discovered that no games could be played, for no two teams consisted of ten different boys. How many boys were in the club?

11. Prove Theorem 3.

12. Prove Theorem 4.

Section 1.7. Subplanes

We wish now to consider one incidence structure contained inside another. We use the language of embeddings and subplanes to do so.

Figure 1.13

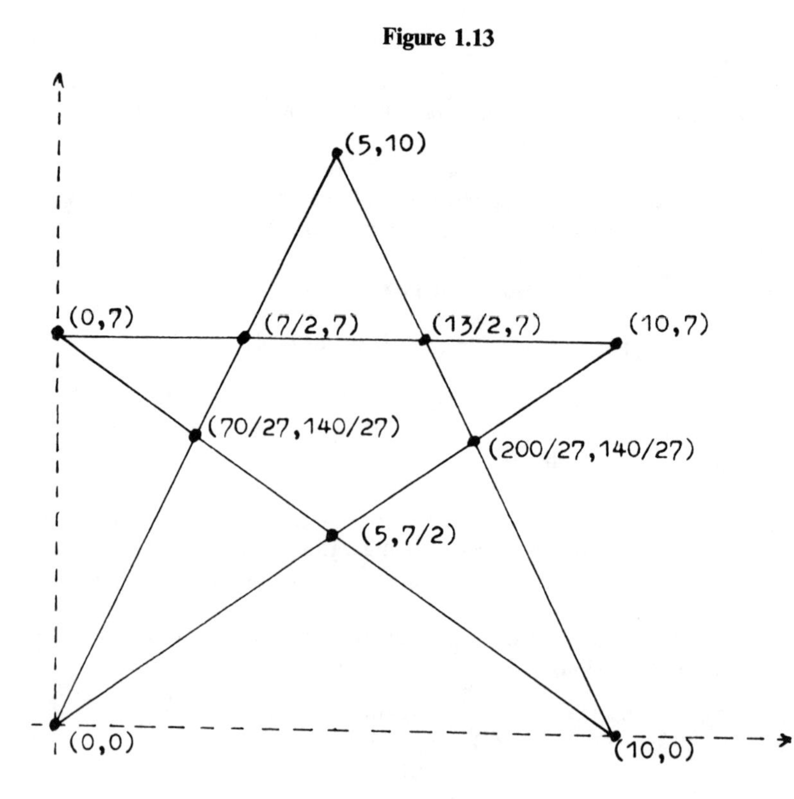

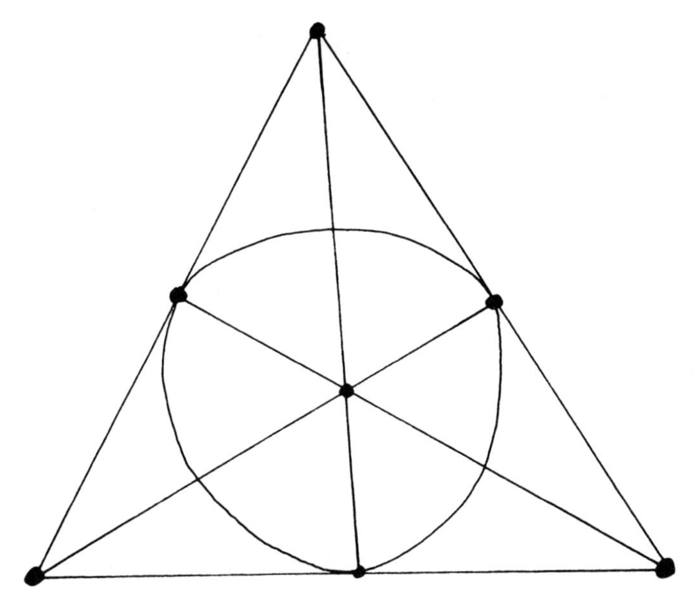

Figure 1.14

Definition 1. Let $\sigma = (\mathcal{P}, \mathcal{L}, \in)$ and $\sigma' = (\mathcal{P}', \mathcal{L}', \in)$ be planes (with lines regarded as sets of points). We say σ is *embedded* in σ' in case

1. $\mathcal{P} \subseteq \mathcal{P}'$ and
2. for every $L \in \mathcal{L}$ there exists $L' \in \mathcal{L}'$ such that $L = L' \cap \mathcal{P}$.

Example 1. Figure 1.13 represents the tactical configurations with form $(10_2, 5_4)$ embedded in the Euclidean plane α_R. To say that the configuration can be embedded in α_R means that it occurs as a subset of α_R that is embedded in α_R. The figure represents one of many possible embeddings.

Example 2. The Fano configuration represented by the Figure 1.14 as drawn is not embedded in α_R, for one of the lines is not a Euclidean line. We shall prove later that the Fano configuration cannot be embedded in α_R; that is, it cannot be drawn with "straight" lines.

Definition 2. Let $\sigma = (\mathcal{P}, \mathcal{L}, \in)$ and $\sigma' = (\mathcal{P}', \mathcal{L}', \in)$ be planes. We say σ is a *subplane* of σ' in case

1. σ is embedded in σ', and
2. if $L' \in \mathcal{L}'$ and $L' \cap \mathcal{P}$ contains two points, then $L' \cap \mathcal{P} \in \mathcal{L}$.

In order for σ to be embedded in σ', each line of σ must be contained in a line of σ'. In order for σ to be a subplane of σ', every line of σ' that contains two points of σ must contain a line of σ.

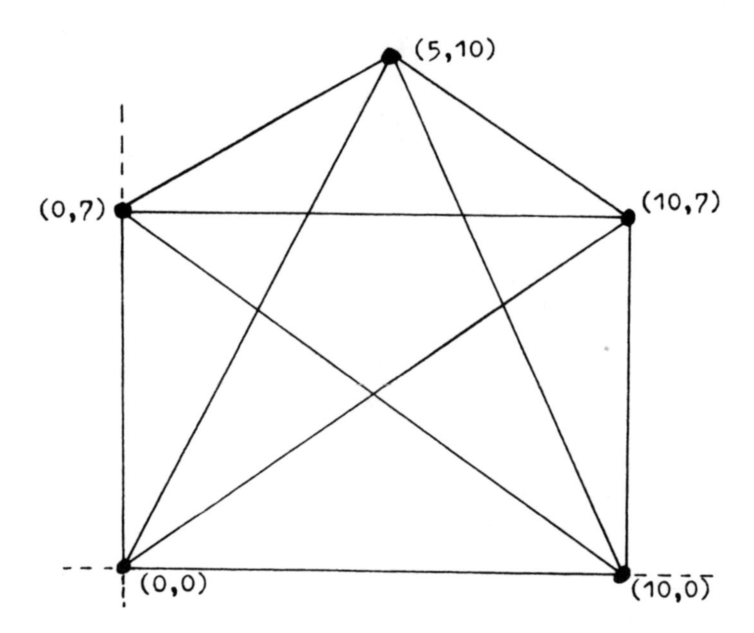

Figure 1.15

Example 3. In the figure of Example 1, the configuration $(10_2, 5_4)$ is embedded in α_R, but is not a subplane of α_R. For example, there is a line in α_R on the two points $(0,0)$ and $(10,0)$, but no line of $(10_2, 5_4)$ is on those two points.

Example 4. Figure 1.15 represents the tactical configuration with form $(5_4, 10_2)$ as a subplane of α_R. It is a subplane because all possible lines belong to the configuration.

Definition 3. If $\sigma = (\mathcal{P}, \mathcal{L}, \in)$ is a subplane of $\sigma' = (\mathcal{P}', \mathcal{L}', \in)$, we say σ is a *principal subplane* of σ' in case $\mathcal{P} = \mathcal{P}' \backslash L'$ for some $L' \in \mathcal{L}'$.

Example 5. A complete four-point is a principal subplane of a Fano configuration. If we use the notation of Examples 1.6.2 and 1.6.3, it is clear that if the line D (and all the points on it) is removed from the Fano configuration, the complete four-point remains. Figure 1.16 illustrates this relationship.

Note, in Example 5, that the Fano configuration, which is a projective plane, has as a principal subplane the complete four-point, which is an affine plane. This sort of thing always happens, as the following theorem indicates.

Theorem 1. *Any principal subplane of a projective plane is an affine plane.*

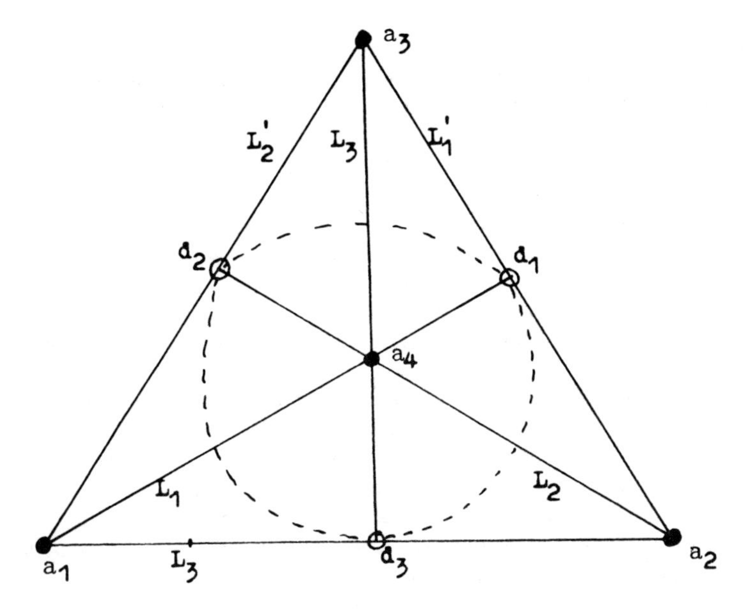

Figure 1.16

PROOF. Let $\pi = (\mathcal{P}', \mathcal{L}', \in)$ be a projective plane, and let L' be a line in π. Let $\mathcal{P} = \mathcal{P}' \backslash L'$, and for each line $M' \in \mathcal{L}'$, $M' \neq L'$, let $M = \{p \in M' \mid p \notin L'\}$. Set $\mathcal{L} = \{M \mid M' \in \mathcal{L}' \backslash \{L'\}\}$. Then let $\alpha = (\mathcal{P}, \mathcal{L}, \in)$; that is, let α be the principal subplane of π obtained by omitting line L'. We shall verify that α is an affine plane by using Exercise 1.2.3.

Suppose that p and q are two points of α. Then there is a unique line $M' \in \mathcal{L}'$ such that $p, q \in M'$. Hence p and q determine the line M in α.

Suppose that M is a line of α, and p is a point of α not on M. Let $q = M' \cap L'$ in π. Then let $N' = pq$ in π. It follows that line N in α is on p, and is parallel to M, for M' and N' meet only at q in π. But q is not a point of α.

In π, there exists a four-point a_0, a_1, a_2, a_3. If L' is on at most one of the points a_0, a_1, a_2, a_3, then any three not on L' are points of α that form a triangle. If L' is on two of the points a_0, a_1, a_2, a_3, say a_2 and a_3, then the point $d_2 = a_0a_2 \cap a_1a_3$ is not on L' or on a_0a_1. Hence a_0, a_1, d_2 constitute a triangle in α.

Thus α is an affine plane. □

The next theorem is a sort of converse of Theorem 1. Not only is the principal subplane of a projective plane an affine plane, but every affine plane arises that way.

Theorem 2. *Every affine plane is a principal subplane of some projective plane.*

PROOF. Let $\alpha = (\mathcal{P}, \mathcal{L}, \in)$ be an affine plane. It is easily verified that the relation "equals or is parallel to" is an equivalence relation on the lines of α. Let p_L denote the parallel class of the line L. We call p_L an *ideal point*. For each line $L \in \mathfrak{L}$, let $L' = L \cup \{p_L\}$, thus extending each line of α to contain an ideal point. Let $L_\infty = \{p_L | L \in \mathfrak{L}\}$, the set of ideal points. We call L_∞ the *ideal line*. Now set $\mathfrak{P}' = \mathfrak{P} \cup L_\infty$ and $\mathfrak{L}' = \{L' | L \in \mathfrak{L}\} \cup \{L_\infty\}$. Let $\pi = (\mathfrak{P}', \mathfrak{L}', \in)$. By construction, α is the principal subplane of π obtained by omitting the line L_∞. It remains only to show that π is a projective plane.

Let p, q be two points of π. If p, q are both points of α, then p and q determine a line M in α; hence p and q determine line M' in π. If p is in α and q is an ideal point, say $q = p_L$, then let M be the line of α which is on p and parallel to L (or L itself if p is on L). Then p and q determine line M' in π. If p and q are both ideal points, then p and q determine line L_∞ in π. Thus two points of π determine a line in π.

Let A and B be two lines in π. If neither A nor B is the ideal line, say $A = L'$ and $B = M'$, consider two cases: If $L \parallel M$, then L' and M' both contain the ideal point p_L; if $L \nparallel M$, then L and M meet in α, so L' and M' meet. If one of A and B is the ideal line, say B, then A and B meet, for A contains an ideal point (as does every line in π). Thus two lines of π always meet.

Finally, α contains a four-point (Exercise 1.2.4), so π contains a four-point. Thus by Exercise 1.2.5, π is a projective plane. □

Thus, according to Theorem 2, any affine plane can be "extended" to a projective plane. It is particularly useful and instructive to extend the real affine plane α_R to a projective plane.

To extend α_R to a projective plane, we first introduce *homogeneous coordinates* in α_R. This is done for points by rewriting the point (x, y) in the form (x_1, x_2, x_3), where $x = x_1/x_3$ and $y = x_2/x_3$. Since

$$\frac{kx_1}{kx_3} = \frac{x_1}{x_3} \quad \text{and} \quad \frac{kx_2}{kx_3} = \frac{x_2}{x_3},$$

the point (kx_1, kx_2, kx_3) is the same as the point (x_1, x_2, x_3). That is, only proportionality counts. It is clear that if (x_1, x_2, x_3) are the homogeneous coordinates of a point in α_R, then $x_3 \neq 0$.

We also introduce homogeneous coordinates for lines by writing the line $ax + by + c = 0$ not as an equation but as a triple $((a, b, c))$. We use the double parentheses simply to distinguish lines from points. Since the equation $kax + kby + kc = 0$ is equivalent to $ax + by + c = 0$, the line $((ka, kb, kc))$ is the same as the line $((a, b, c))$. That is, only proportionality counts for line coordinates, too. It is the case that if $((a, b, c))$ is the set of homogeneous coordinates of a line in α_R, then a and b are not both zero.

Now, we ask, in terms of homogeneous coordinates, what is the incidence condition? The answer is that (x_1, x_2, x_3) is on $((a, b, c))$ if and only if $(x_1/x_3, x_2/x_3)$ is on $ax + by + c = 0$, if and only if $a(x_1/x_3) + b(x_2/x_3) + c = 0$, if and only if $ax_1 + bx_2 + cx_3 = 0$. We shall use this last equation as the incidence condition.

Now that we have a homogeneous coordinate system on α_R, we can begin the extension process. First note that if $((a, b, c))$ and $((a', b', c'))$ are two lines in α_R, the point of intersection is $(bc' - b'c, ca' - c'a, ab' - a'b)$—that is, if the two lines are not parallel. But if they are parallel, then $a = a'$ and $b = b'$, with $c \neq c'$, and the point-of-intersection formula becomes $(bc' - bc, ca - c'a, ab - ab) = (b(c' - c), a(c - c'), 0) = (b, -a, 0)$, not a point of α_R, of course. But we may adopt $(b, -a, 0)$ as an ideal point. Thus an ideal point is any set of coordinates of the form $(x_1, x_2, 0)$.

Next, we look for some representation of the ideal line. We discover that the triple $((0,0,1))$ satisfies the incidence condition for every ideal point, so we agree to call $((0,0,1))$ the ideal line.

Now the extension of α_R we have obtained consists of points (x_1, x_2, x_3) with proportional triples representing the same point, the point being ideal in case $x_3 = 0$, and of lines $((a,b,c))$, with proportional triples representing the same line, the line being ideal in case $a = b = 0$; the incidence condition is $ax_1 + bx_2 + cx_3 = 0$. Note that the triple $((0,0,0))$ does not represent any line. Neither does the triple $(0,0,0)$ represent any point, for it is not a point of α_R, and if it were the ideal point on line $((a,b,c))$, it would have to be proportional to $(b, -a, 0)$, which it is not.

It should be abundantly clear now that the extension of α_R we have obtained is isomorphic to π_R, under the mappings $f:(x_1, x_2, x_3) \rightarrow [x_1, x_2, x_3]$, $F:((a,b,c)) \rightarrow \langle a,b,c \rangle$. We thus have the following theorem.

Theorem 3. *α_R is a principal subplane of π_R.*

The reverse process, that of restricting π_R to obtain a principal subplane, yields the same result. If we remove the line $\langle 0,0,1 \rangle$ from π_R, together with the points on it, which are all of the form $[x_1, x_2, 0]$, we obtain a principal subplane in which each point can be written in the form $[x_1, x_2, 1]$ and in which each line $\langle l_1, l_2, l_3 \rangle$ has the property that l_1 and l_2 are not both zero. Hence this subplane is isomorphic to α_R under the mappings $f:(x, y) \rightarrow [x, y, 1]$, $F: ax + by + c = 0 \rightarrow \langle a,b,c \rangle$.

The fact that α_R is a subplane of π_R under the isomorphisms displayed means that results in π_R may be applied to α_R. For example, the point of intersection of lines $ax + by + c = 0$ and $a'x + b'y + c' = 0$ in α_R is the same as the point on both $\langle a,b,c \rangle$ and $\langle a',b',c' \rangle$ in π_R, so Theorem 1.3.2 applies, giving us a quick and easy way to solve two linear equations in two unknowns simultaneously. We shall have the opportunity to say several things about α_R as a result of Theorem 3.

Starting with a line $ax + by + c = 0$ in α_R, we find the ideal point on the line by going to its homogeneous form, $ax_1 + bx_2 + cx_3 = 0$, and then requiring that $x_3 = 0$. That gives us the equation $ax_1 + bx_2 = 0$, so that $x_1 = -(b/a)x_2$. Hence the ideal point on the original line is $[x_1, x_2, x_3] = [-(b/a)x_2, x_2, 0] = [-b, a, 0]$.

The same process is used to find the ideal points on any locus in α_R. We start with the equation of the locus, $f(x, y) = 0$, and generate its homogeneous form by substituting $x = x_1/x_3, y = x_2/x_3$, and then multiplying by x_3 a sufficient number of times to clear the denominators. (This only works for algebraic functions f, of course; we shall not consider any but polynomial functions.) Then we solve the resulting equation simultaneously with $x_3 = 0$, and we obtain the ideal points.

We sometimes call the Euclidean locus, together with the ideal points on it, the *extension* of the original locus in α_R.

Example 6. The parabola $y = x^2$ has the homogeneous form $x_2x_3 = x_1^2$, or $x_1^2 - x_2x_3 = 0$. Requiring $x_3 = 0$, we find $x_1 = 0$, so the point $[0, x_2, 0] = [0, 1, 0]$ is an ideal point (the only one) on the extension of the parabola.

Exercises 1.7

1. Prove that a complete four-point is a subplane of every affine plane.

2. Show that a Pappus configuration can be embedded in π_R.

3. Show that a Desargues configuration can be embedded in π_R.

4. Give the homogeneous form of each of the equations.

   (a) $x + 3y = 5$, **(d)** $x^2 + y^2 = 1$,
   (b) $2x - y + 3 = 0$, **(e)** $xy + x^2 - 2 = 0$,
   (c) $xy = 1$, **(f)** $x^2 + 2xy + y^2 - 3x + y + 4 = 0$.

5. Find the ideal points that belong to loci defined by the equations in Exercise 4.

6. Prove that the conic section $ax^2 + bxy + cy^2 + dx + ey + f = 0$ is an ellipse, parabola, or hyperbola according as its extension meets the ideal line in 0, 1, or 2 points.

7. Show that a line parallel to the axis of the parabola $y = ax^2 + bx + c$ meets the parabola in one point of α_R and in one ideal point.

8. Show that a line parallel to an asymptote of the hyperbola $x^2/a^2 - y^2/b^2 = 1$ meets the hyperbola in one point of α_R and in one ideal point.

Section 1.8. Further Examples

In this section, we look at some more examples of projective planes, and consider some existence theorems.

Example 1. Let D be a division ring. Let $T = D \times D \times D \setminus \{(0,0,0)\}$, the set of all nonzero triples of D. We say triples (x_1, x_2, x_3) and (y_1, y_2, y_3) in T are *left proportional* in case there exists $k \in D$ such that $x_i = ky_i$, $i = 1, 2, 3$. Left proportionality is an equivalence relation; let $[x_1, x_2, x_3]$ be the left-proportionality class of (x_1, x_2, x_3) in α. Similarly, let $\langle l_1, l_2, l_3\rangle$ be the right-proportionality class of (l_1, l_2, l_3) in T. Let

$$\mathcal{P}_D = \{[x_1, x_2, x_3] \mid (x_1, x_2, x_3) \in T\},$$
$$\mathcal{L}_D = \{\langle l_1, l_2, l_3\rangle \mid (l_1, l_2, l_3) \in T\},$$
$$\mathcal{I}_D = \{([x_1, x_2, x_3], \langle l_1, l_2, l_3\rangle) \mid x_1 l_1 + x_2 l_2 + x_3 l_3 = 0\}.$$

The incidence structure $\pi_D = (\mathcal{P}_D, \mathcal{L}_D, \mathcal{I}_D)$ is called the *plane over D*. π_D is a projective plane.

***Example** 2*. Let F be a field. Then F is a division ring, so π_F may be defined as in Example 1. The plane π_R is of this form.

***Example** 3*. Let C be the field of complex numbers. The plane π_C is called the *complex projective plane* or the *classical projective plane*.

Theorem 1. *If K is a subfield of field F, then π_K is a subplane of π_F.*

The proof of Theorem 1 is left as an exercise. As a consequence of Theorem 1, we have π_R a subplane of π_C. Thus α_R is also a subplane of π_C.

We should remark here that Theorems 1.3.1, 1.3.2, 1.3.3, and 1.3.4 all hold in π_F for F any field. Theorem 1.3.4 also holds in π_D for D any division ring.

***Example** 4*. Let $V = \{0, 1, 2, i, 2i, j, 2j, k, 2k\}$, and let $+$ and $\cdot$ be defined on V as in the following tables:

$+$	0	1	2	i	$2i$	j	$2j$	k	$2k$
0	0	1	2	i	$2i$	j	$2j$	k	$2k$
1	1	2	0	$2k$	j	k	i	$2i$	$2j$
2	2	0	1	$2j$	k	$2i$	$2k$	j	i
i	i	$2k$	$2j$	$2i$	0	1	k	2	j
$2i$	$2i$	j	k	0	i	$2k$	2	$2j$	1
j	j	k	$2i$	1	$2k$	$2j$	0	i	2
$2j$	$2j$	i	$2k$	k	2	0	j	1	$2i$
k	k	$2i$	j	2	$2j$	i	1	$2k$	0
$2k$	$2k$	$2j$	i	j	1	2	$2i$	0	k

$\cdot$	0	1	2	i	$2i$	j	$2j$	k	$2k$
0	0	0	0	0	0	0	0	0	0
1	0	1	2	i	$2i$	j	$2j$	k	$2k$
2	0	2	1	$2i$	i	$2j$	j	$2k$	k
i	0	i	$2i$	2	1	k	$2k$	$2j$	j
$2i$	0	$2i$	i	1	2	$2k$	k	j	$2j$
j	0	j	$2j$	$2k$	k	2	1	i	$2i$
$2j$	0	$2j$	j	k	$2k$	1	2	$2i$	i
k	0	k	$2k$	j	$2j$	$2i$	i	2	1
$2k$	0	$2k$	k	$2j$	j	i	$2i$	1	2

The structure $V, +$ is the additive (abelian) group of a field with nine elements; the structure $V \backslash \{0\}, \cdot$ is the so-called quaternion group of order 8. It can be verified that $V, +, \cdot$ satisfies the right distributive law $(a+b)c = ac + bc$, but not the left distributive law $a(b+c) = ab + ac$. The structure $V, +, \cdot$ is called a *right nearfield*. The *plane over* V is the incidence structure $\pi_V = (\mathcal{P}, \mathcal{L}, \mathcal{I})$ with

$$\mathcal{P} = \{[x_1, x_2, 1] \mid x_1, x_2 \in V\} \cup \{[1, x_2, 0] \mid x_2 \in V\} \cup \{[0,1,0]\},$$
$$\mathcal{L} = \{\langle l_1, 1, l_2 \rangle \mid l_1, l_3 \in V\} \cup \{\langle 1, 0, l_3 \rangle \mid l_3 \in V\} \cup \{\langle 0,0,1 \rangle\},$$
$$\mathcal{I} = \{([x_1, x_2, x_3], \langle l_1, l_2, l_3 \rangle) \mid x_1 l_1 + x_2 l_2 + x_3 l_3 = 0\}.$$

It can now be verified that π_V is a projective plane, one of a class of planes called *Veblen–Wedderburn planes*. The verification is left to the exercises.

***Example** 5*. Let V be as in Example 4. Define $\odot$ on V by $x \odot y = y \cdot x$. The structure $V, +, \odot$ is called the *dual structure* of $V, +, \cdot$, and is denoted by V^d. V^d is also a left nearfield: the left distributive law holds, but not the right distributive law. The plane over V^d can be defined by dualizing the definition of p_V, and $\pi_{V^d} = \pi_V^d$. It can be shown (it is beyond our scope to do so) that π_V and π_V^d are *not* isomorphic.

We close this section with a brief discussion of the existence of projective planes. To facilitate the discussion we introduce the concept of the order of a plane.

Definition 1. Let α be a finite affine plane. The *order* of α, denoted by $o(\alpha)$, is the number of points on any line of α.

Referring to Theorem 1.6.4, we see that α has the form $(n^2_{n+1}, n^2 + n_n)$, where $n = o(\alpha)$.

Definition 2. Let π be a finite projective plane. The *order* of π, denoted $o(\pi)$, is the order of any principal subplane of π.

Since a principal subplane of π is an affine plane, the definition of $o(\pi)$ makes sense. Using Theorems 1.6.2 and 1.6.4, you can show that if $o(\pi) = n$, then π has the form $(n^2 + n + 1_{n+1})$.

We state the following theorem as an aid to answering the existence questions.

Theorem 2. *If F is a finite field with q elements, then $o(\pi_F) = q$.*

PROOF. Points of F are represented by triples $[x_1, x_2, x_3]$ in which each $x_i \in F$, not all the x_i are zero, and proportional triples represent the same points. We will begin counting the points of π_F by counting triples. For each x_i, there are q elements of F that may be chosen, so there are q^3 triples in all, including the zero triple. Hence $q^3 - 1$ triples may represent points. Now we find the number of points by dividing by the number of triples proportional to a given triple. Starting with a given triple, we may select any nonzero element of F with which to multiply each element of the triple, in order to obtain a proportional triple. Hence there are $q-1$ triples proportional to a given triple. Thus we see there are

$$\frac{q^3-1}{q-1} = q^2+q+1$$

points in π_F. If $o(\pi_F) = n$, then π_F has n^2+n+1 points; hence

$$n^2+n+1 = q^2+q+1.$$

Factoring, we get

$$(n-q)(n+q+1) = 0,$$

whose only positive root is $n = q$. Thus $o(\pi_F) = q$. □

We can now state the existence theorems.

Theorem 3. *If q is a power of a prime, then there exists a projective plane of order q.*

The proof of Theorem 3 is based on the fact that if q is a power of a prime, then there exists a field with q elements. See Veblen and Bussey [17]; further algebraic details can be found in Herstein [11].

Theorem 4. *If q is congruent to 1 or 2 modulo 4 and is not the sum of the squares of two integers, then there exists no projective plane of order q.*

Theorem 4 was proved by Bruck and Ryser [4] in 1949. A proof is also found in Hughes and Piper [12].

Exercises 1.8

1. Verify in detail that π_D is a projective plane.

2. Show that if Z_2 is the field of integers modulo 2, then π_{Z_2} is isomorphic to the Fano configuration.

3. Let the equation $f(x, y) = 0$ define a locus in α_R. The *extension* of the locus in π_C is the set of all points in π_C satisfying the homogeneous form of the equation $f(x, y) = 0$. Show that the conic $ax^2 + bxy + cy^2 + dx + ey + f = 0$ in α_R is a circle if and only if its extension in π_C contains the two points $[1, i, 0]$ and $[1, -i, 0]$.

4. Prove Theorem 1.

5. Verify that π_V is a plane.

6. Show that π_V is a projective plane by using Theorem 1.6.1.

7. Write out the definition of $\pi_V^d = \pi_{V^d}$.

8. Is every projective plane self-dual?

9. Is every tactical configuration with form (m_r) self-dual?

10. If π is a finite projective plane with form $(n^2 - n + 1_n)$, and α is a principal subplane of π, find the form of α.

11. If α is a finite affine plane with form $(n^2_{n+1}, n^2 + n_n)$, and α is a principal subplane of projective plane π, find the form of π.

12. Determine all positive integers $n \leqslant 50$ for which there exist projective planes of order n, according to Theorem 3.

13. Determine all positive integers $n \leqslant 50$ for which there do not exist projective planes of order n, according to the Bruck–Ryser theorem.

14. Does there exist a projective plane of order 10?

***15.** Let D be a division ring, and let $x = [x_1, x_2, x_3]$, $y = [y_1, y_2, y_3]$, and $z = [z_1, z_2, z_3]$ be three points in π_D. Show that x, y, and z are collinear if and only if there exist $\alpha, \beta, \gamma \in D$, not all zero, such that $\alpha x_i + \beta y_i + \gamma z_i = 0$, $i = 1, 2, 3$.

Chapter 2

Collineations

One of the most fruitful means of studying mathematical structures has been through the use of functions. This is particularly true in such algebraic topics as group theory and ring theory, where homomorphisms are a powerful tool of investigation.

It is no less true that certain functions have proved to be valuable tools of investigation in projective geometry. The functions are called collineations, and are just isomorphisms from a projective plane to itself. In this chapter, we shall develop the theory of collineations, and indicate some of their applications.

Section 2.1. Perspectivities

Before we consider a function from a projective plane to itself, we need to consider certain functions from one range to another in the same plane.

Let L and L' be two ranges in projective plane π, and let u be a point not on L or L'. The function $f: L \to L'$ defined by

$$f(x) = L' \cap ux$$

is called a *perspectivity* from L to L' with *center* u. We use the notation

$$f: L \overset{u}{\barwedge} L'$$

to indicate that f is a perspectivity from L to L' with center u. The notation

$$f: L \barwedge L'$$

indicates that f is a perspectivity from L to L', but does not name the center. If $a, b, c, \ldots$ are points on L and $a', b', c', \ldots$ are points on L', the notation

$$f : L(a,b,c,\ldots) \overline{\wedge} L'(a',b',c',\ldots)$$

indicates that f is a perspectivity from L to L', and $a' = f(a)$, $b' = f(b)$, $c' = f(c), \ldots$. Note that in this case the lines $aa', bb', cc', \ldots$ must all be on the center of the perspectivity.

Some immediate consequences of the definition of perspectivity are summarized in the next theorem, whose proof is left as an exercise.

Theorem 1. *A perspectivity from L to L′ is a bijection, and its inverse is a perspectivity from L′ to L, having the same center.*

We shall investigate the properties of perspectivities more thoroughly in the plane π_F over a field F. We will obtain some algebraic characterizations that will be interesting and useful in their own right, and will also serve to illustrate the theory.

Let L and L' be two lines (ranges) in π_F. Let $a = [a_1, a_2, a_3]$ and $b = [b_1, b_2, b_3]$ be two points on L, which we shall use as base points in a parametrization of L (see Section 1.3). Let $a' = [a'_1, a'_2, a'_3]$ and $b' = [b'_1, b'_2, b'_3]$ be base points on L'. Let $u = [u_1, u_2, u_3]$ be a point not on L or L', and let $f : L \overset{u}{\overline{\wedge}} L'$ be the perspectivity from L to L' with center u.

Note that it might or might not be the case that $f(a) = a'$, according as u is or is not on line aa'.

For any point c on L, let c have parameters (λ, μ) relative to a and b. Then $f(c)$ is on L'; let $f(c)$ have parameters (λ', μ') relative to a' and b'.

We can now derive an equation relating the parameters (λ', μ') to the parameters (λ, μ) that is characteristic of the perspectivity f. Let

$$\alpha = \begin{vmatrix} u_1 & u_2 & u_3 \\ a_1 & a_2 & a_3 \\ a'_1 & a'_2 & a'_3 \end{vmatrix}, \qquad \beta = \begin{vmatrix} u_1 & u_2 & u_3 \\ a_1 & a_2 & a_3 \\ b'_1 & b'_2 & b'_3 \end{vmatrix},$$

$$\gamma = \begin{vmatrix} u_1 & u_2 & u_3 \\ b_1 & b_2 & b_3 \\ a'_1 & a'_2 & a'_3 \end{vmatrix}, \qquad \delta = \begin{vmatrix} u_1 & u_2 & u_3 \\ b_1 & b_2 & b_3 \\ b'_1 & b'_2 & b'_3 \end{vmatrix}.$$

Theorem 2. $\alpha\lambda\lambda' + \beta\lambda\mu' + \gamma\mu\lambda' + \delta\mu\mu' = 0.$

PROOF. Since the points u, $c = [\lambda a_1 + \mu b_1, \lambda a_2 + \mu b_2, \lambda a_3 + \mu b_3]$, and $f(c) = [\lambda' a'_1 + \mu' b'_1, \lambda' a'_2 + \mu' b'_2, \lambda' a'_3 + \mu' b'_3]$ are collinear, we have by Theorem 1.3.3

$$\begin{vmatrix} u_1 & u_2 & u_3 \\ \lambda a_1 + \mu b_1 & \lambda a_2 + \mu b_2 & \lambda a_3 + \mu b_3 \\ \lambda' a'_1 + \mu' b'_1 & \lambda' a'_2 + \mu' b'_2 & \lambda' a'_3 + \mu' b'_3 \end{vmatrix} = 0.$$

The determinant can be expanded to

$$\alpha\lambda\lambda' + \beta\lambda\mu' + \gamma\mu\lambda' + \delta\mu\mu',$$

and the theorem follows. □

The equation in Theorem 2 is called the *equation of the perspectivity*. It is obviously the case that the equation of f depends upon the parametrizations of L and L', for the coefficients in the equation depend on the base points chosen. That is, if the parametrizations change, the equation will change, even though the perspectivity does not change.

Given base points for L and L' and a center u, the equation of the perspectivity $f: L \overset{u}{\barwedge} L'$ can be found simply by computing α, β, γ, and δ. On the other hand, given the equation and the base points for L and L', we can find the center and we can find the image of any given point on L.

Example 1. Let line L have base points $a = [1, 2, -1]$ and $b = [0, 1, 1]$, and let line L' have base points $a' = [1, 0, 1]$ and $b' = [0, 1, -1]$. Suppose perspectivity $f: L \barwedge L'$ has equation

$$2\lambda\lambda' - \lambda\mu' + 3\mu\mu' = 0.$$

Then $\alpha = 2$, $\beta = -1$, $\gamma = 0$, $\delta = 3$. We can find the center $u = [u_1, u_2, u_3]$ from the equations

$$\alpha = \begin{vmatrix} u_1 & u_2 & u_3 \\ 1 & 2 & -1 \\ 1 & 0 & 1 \end{vmatrix} = 2u_1 - 2u_2 - 2u_3 = 2,$$

$$\beta = \begin{vmatrix} u_1 & u_2 & u_3 \\ 1 & 2 & -1 \\ 0 & 1 & -1 \end{vmatrix} = -u_1 + u_2 + u_3 = -1,$$

$$\gamma = \begin{vmatrix} u_1 & u_2 & u_3 \\ 0 & 1 & 1 \\ 1 & 0 & 1 \end{vmatrix} = u_1 + u_2 - u_3 = 0,$$

$$\delta = \begin{vmatrix} u_1 & u_2 & u_3 \\ 0 & 1 & 1 \\ 0 & 1 & -1 \end{vmatrix} = -2u_1 = 3;$$

we get

$$u = \left[-\tfrac{3}{2}, -\tfrac{1}{2}, -2\right] = [3, 1, 4].$$

To find images of points on L, we use the equation directly. For example, point a has parameters $(1, 0)$; hence the parameters (λ', μ') of $f(a)$ satisfy

$$2(1)\lambda' - (1)\mu' + 3(0)\mu' = 0,$$

$$2\lambda' = \mu',$$

$$(\lambda', \mu') = (1, 2).$$

Hence $f(a)$ is the point $[1(1)+2(0), 1(0)+2(1), 1(1)+2(-1)] = [1,2,-1]$, which just happens to be a itself. That is, L and L' meet at a. The point b has parameters $(0, 1)$, so its image $f(b)$ has parameters satisfying

$$2(0)\lambda' - (0)\mu' + 3(1)\mu' = 0,$$
$$\mu' = 0,$$
$$(\lambda', \mu') = (1,0).$$

Hence $f(b) = a' = [1,0,1]$. The point $[1,3,0]$ on L has parameters $(1, 1)$, so its image has parameters satisfying

$$2(1)\lambda' - (1)\mu' + 3(1)\mu' = 0,$$
$$2\lambda' + 2\mu' = 0,$$
$$\lambda' = -\mu'$$
$$(\lambda', \mu') = (1, -1).$$

Hence $f([1,3,0]) = [1, -1, 2]$.

The form of the equation of a perspectivity given in Theorem 2 is not always the most convenient. An equivalent matrix form is often easier to use, as we shall see.

Theorem 3. *The equation*

$$\alpha\lambda\lambda' + \beta\lambda\mu' + \gamma\mu\lambda' + \delta\mu\mu' = 0$$

is equivalent to the matrix equation

$$(\lambda', \mu') = (\lambda, \mu)\begin{pmatrix} \beta & -\alpha \\ \delta & -\gamma \end{pmatrix}.$$

PROOF. $\alpha\lambda\lambda' + \beta\lambda\mu' + \gamma\mu\lambda' + \delta\mu\mu' = 0$ if and only if

$$\beta\lambda\mu' + \delta\mu\mu' = -\alpha\lambda\lambda' - \gamma\mu\lambda',$$

if and only if

$$(\beta\lambda + \delta\mu)\mu' = (-\alpha\lambda - \gamma\mu)\lambda',$$

if and only if

$$\lambda' = \beta\lambda + \delta\mu,$$
$$\mu' = -\alpha\lambda - \gamma\mu,$$

if and only if

$$(\lambda', \mu') = (\lambda, \mu)\begin{pmatrix} \beta & -\alpha \\ \delta & -\gamma \end{pmatrix}.$$

□

Combining Theorems 2 and 3, we see that any perspectivity f in π_F has an equation of the form

$$f(\lambda, \mu) = (\lambda, \mu)M,$$

in terms of parameters, with M a 2×2 matrix. (The fact that M is nonsingular follows from the fact that f is a bijection, and is left as an exercise.) The question arises whether every equation of this form defines a perspectivity. The answer is no, as is fully amplified by the following theorem.

Theorem 4. *Let L and L' be two lines in π_F, L having base points $a=[a_1,a_2,a_3]$ and $b=[b_1,b_2,b_3]$, and L' having base points $a'=[a_1',a_2',a_3']$ and $b'=[b_1',b_2',b_3']$. Let*

$$p_1=\begin{vmatrix} \begin{vmatrix} a_2 & a_3 \\ b_2' & b_3' \end{vmatrix} & \begin{vmatrix} a_3 & a_1 \\ b_3' & b_1' \end{vmatrix} & \begin{vmatrix} a_1 & a_2 \\ b_1' & b_2' \end{vmatrix} \\ \begin{vmatrix} b_2 & b_3 \\ a_2' & a_3' \end{vmatrix} & \begin{vmatrix} b_3 & b_1 \\ a_3' & a_1' \end{vmatrix} & \begin{vmatrix} b_1 & b_2 \\ a_1' & a_2' \end{vmatrix} \\ \begin{vmatrix} b_2 & b_3 \\ b_2' & b_3' \end{vmatrix} & \begin{vmatrix} b_3 & b_1 \\ b_3' & b_1' \end{vmatrix} & \begin{vmatrix} b_1 & b_2 \\ b_1' & b_2' \end{vmatrix} \end{vmatrix},$$

$$p_2=\begin{vmatrix} \begin{vmatrix} a_2 & a_3 \\ a_2' & a_3' \end{vmatrix} & \begin{vmatrix} a_3 & a_1 \\ a_3' & a_1' \end{vmatrix} & \begin{vmatrix} a_1 & a_2 \\ a_1' & a_2' \end{vmatrix} \\ \begin{vmatrix} b_2 & b_3 \\ a_2' & a_3' \end{vmatrix} & \begin{vmatrix} b_3 & b_1 \\ a_3' & a_1' \end{vmatrix} & \begin{vmatrix} b_1 & b_2 \\ a_1' & a_2' \end{vmatrix} \\ \begin{vmatrix} b_2 & b_3 \\ b_2' & b_3' \end{vmatrix} & \begin{vmatrix} b_3 & b_1 \\ b_3' & b_1' \end{vmatrix} & \begin{vmatrix} b_1 & b_2 \\ b_1' & b_2' \end{vmatrix} \end{vmatrix},$$

$$p_3=\begin{vmatrix} \begin{vmatrix} a_2 & a_3 \\ a_2' & a_3' \end{vmatrix} & \begin{vmatrix} a_3 & a_1 \\ a_3' & a_1' \end{vmatrix} & \begin{vmatrix} a_1 & a_2 \\ a_1' & a_2' \end{vmatrix} \\ \begin{vmatrix} a_2 & a_3 \\ b_2' & b_3' \end{vmatrix} & \begin{vmatrix} a_3 & a_1 \\ b_3' & b_1' \end{vmatrix} & \begin{vmatrix} a_1 & a_2 \\ b_1' & b_2' \end{vmatrix} \\ \begin{vmatrix} b_2 & b_3 \\ b_2' & b_3' \end{vmatrix} & \begin{vmatrix} b_3 & b_1 \\ b_3' & b_1' \end{vmatrix} & \begin{vmatrix} b_1 & b_2 \\ b_1' & b_2' \end{vmatrix} \end{vmatrix},$$

$$p_4=\begin{vmatrix} \begin{vmatrix} a_2 & a_3 \\ a_2' & a_3' \end{vmatrix} & \begin{vmatrix} a_3 & a_1 \\ a_3' & a_1' \end{vmatrix} & \begin{vmatrix} a_1 & a_2 \\ a_1' & a_2' \end{vmatrix} \\ \begin{vmatrix} a_2 & a_3 \\ b_2' & b_3' \end{vmatrix} & \begin{vmatrix} a_3 & a_1 \\ b_3' & b_1' \end{vmatrix} & \begin{vmatrix} a_1 & a_2 \\ b_1' & b_2' \end{vmatrix} \\ \begin{vmatrix} b_2 & b_3 \\ a_2' & a_3' \end{vmatrix} & \begin{vmatrix} b_3 & b_1 \\ a_3' & a_1' \end{vmatrix} & \begin{vmatrix} b_1 & b_2 \\ a_1' & a_2' \end{vmatrix} \end{vmatrix}.$$

Then the function $f\colon L\to L'$, defined in terms of parameters by

$$f(\lambda,\mu)=(\lambda,\mu)\begin{pmatrix} \beta & -\alpha \\ \delta & -\gamma \end{pmatrix}, \qquad \alpha\delta-\beta\gamma\neq 0,$$

is a perspectivity if and only if

$$p_1\alpha - p_2\beta + p_3\gamma - p_4\delta = 0.$$

PROOF. From the definitions of $\alpha, \beta, \gamma, \delta$ preceding Theorem 2, we can derive the system of equations

$$\begin{vmatrix} a_2 & a_3 \\ a_2' & a_3' \end{vmatrix} u_1 + \begin{vmatrix} a_3 & a_1 \\ a_3' & a_1' \end{vmatrix} u_2 + \begin{vmatrix} a_1 & a_2 \\ a_1' & a_2' \end{vmatrix} u_3 + \alpha(-1) = 0,$$

$$\begin{vmatrix} a_2 & a_3 \\ b_2' & b_3' \end{vmatrix} u_1 + \begin{vmatrix} a_3 & a_1 \\ b_3' & b_1' \end{vmatrix} u_2 + \begin{vmatrix} a_1 & a_2 \\ b_1' & b_2' \end{vmatrix} u_3 + \beta(-1) = 0,$$

$$\begin{vmatrix} b_2 & b_3 \\ a_2' & a_3' \end{vmatrix} u_1 + \begin{vmatrix} b_3 & b_1 \\ a_3' & a_1' \end{vmatrix} u_2 + \begin{vmatrix} b_1 & b_2 \\ a_1' & a_2' \end{vmatrix} u_3 + \gamma(-1) = 0,$$

$$\begin{vmatrix} b_2 & b_3 \\ b_2' & b_3' \end{vmatrix} u_1 + \begin{vmatrix} b_3 & b_1 \\ b_3' & b_1' \end{vmatrix} u_2 + \begin{vmatrix} b_1 & b_2 \\ b_1' & b_2' \end{vmatrix} u_3 + \delta(-1) = 0$$

in the "unknowns" u_1, u_2, u_3, and -1. If f is a perspectivity, then it has a center u, so the system above has a solution. Hence the determinant of coefficients is zero; when expanded by the last column, the determinant yields

$$p_1\alpha - p_2\beta + p_3\gamma - p_4\delta = 0.$$

Conversely, if this equation holds, then the determinant of coefficients of the above system is zero. Hence the system has a nontrivial solution for u_1, u_2, u_3. That is, if $u_1 = u_2 = u_3 = 0$, then $\alpha = \beta = \gamma = \delta = 0$, so $\alpha\delta - \beta\gamma = 0$, a contradiction. Moreover, the equation

$$f(\lambda, \mu) = (\lambda, \mu)\begin{pmatrix} \beta & -\alpha \\ \delta & -\gamma \end{pmatrix}$$

defining f can be written $\alpha\lambda\lambda' + \beta\lambda\mu' + \gamma\mu\lambda' + \delta\mu\mu' = 0$, which can be written

$$\begin{vmatrix} u_1 & u_2 & u_3 \\ \lambda a_1 + \mu b_1 & \lambda a_2 + \mu b_2 & \lambda a_3 + \mu b_3 \\ \lambda' a_1' + \mu' b_1' & \lambda' a_2' + \mu' b_2' & \lambda' a_3' + \mu' b_3' \end{vmatrix} = 0.$$

Hence the points u, (λ, μ), and $f(\lambda, \mu)$ are collinear, and f is a perspectivity. □

Exercises 2.1

1. Show that a perspectivity is uniquely determined by two points and their images.

2. If $(\lambda', \mu') = (\lambda, \mu)M$ is the equation of a perspectivity, show that the 2×2 matrix M is nonsingular.

3. Let line L in π_R have base points $a = [1, 1, 0]$ and $b = [0, 0, 1]$, and let line L' have base points $a' = [1, 2, 0]$ and $b' = [1, 0, -1]$. If $u = [1, 0, 0]$, find the equation of the perspectivity $f\colon L \overset{u}{\barwedge} L'$.

4. Rework Problem 3 with $a' = [0, 2, 1]$ and $b' = [2, -2, -3]$.

5. Let L, L', a, b, a', b' be as in Problem 3. If the perspectivity $f: L \overset{v}{\barwedge} L'$ has the equation

$$(\lambda', \mu') = (\lambda, \mu)\begin{pmatrix} 1 & -1 \\ 0 & -4 \end{pmatrix},$$

find the coordinates of point v.

6. In π_R, let $L_1 = \langle 1, 0, 0 \rangle$, $L_2 = \langle 0, 1, 0 \rangle$, $L_3 = \langle 0, 0, 1 \rangle$, $u = [1, -1, 1]$, $v = [1, 1, 1]$, and $w = [1, 1, -1]$. Let $f: L_1 \overset{u}{\barwedge} L_2$, $g: L_2 \overset{v}{\barwedge} L_3$, $h: L_3 \overset{w}{\barwedge} L_1$. Show that, for any point p_1 of L_1, $hgf(p_1) = p_1$.

Section 2.2. Projectivities

Having studied perspectivities between ranges in a projective plane, we now study a generalization of that idea. The diagram in Figure 2.1 suggests that the composition of two perspectivities,

$$L_1(a_1b_1c_1) \overset{u_1}{\barwedge} L_2(a_2b_2c_2) \overset{u_2}{\barwedge} L_3(a_3b_3c_3),$$

Figure 2.1

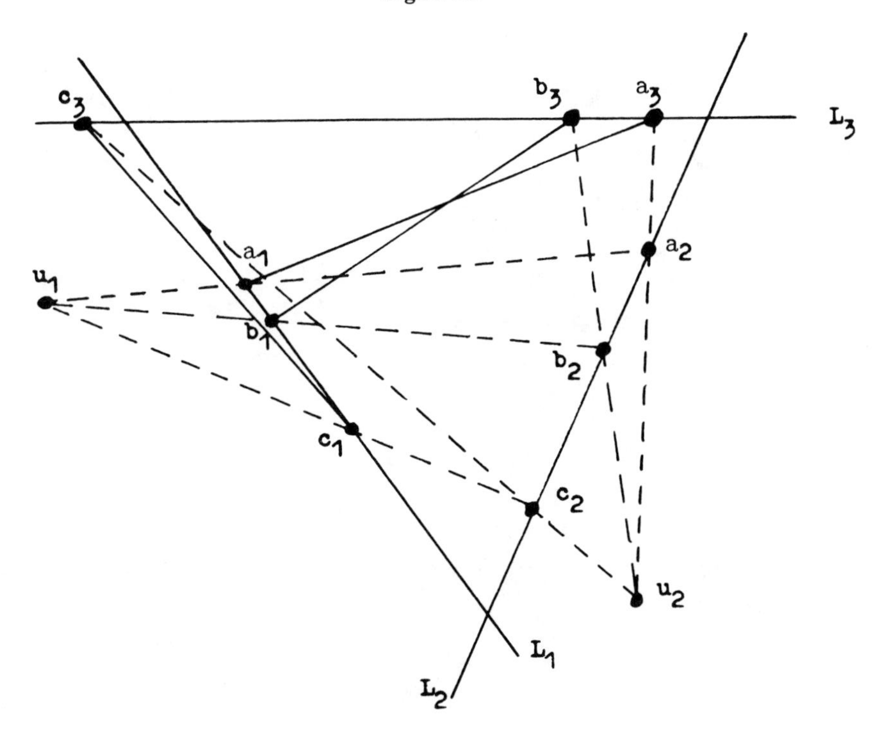

need not be a perspectivity, for lines a_1a_3, b_1b_3, and c_1c_3 need not be concurrent. Yet the function from L_1 to L_3 carrying a_1 to a_3, b_1 to b_3, and c_1 to c_3 ought to be of interest; it is an example of a projectivity.

A *projectivity* is the composition of a finite number of perspectivities. If $f: L \to L'$ is a projectivity, then either f is itself a perspectivity or there must exist lines $L_1, L_2, \ldots, L_n$ such that

$$f : L \barwedge L_1 \barwedge L_2 \barwedge \cdots \barwedge L_n \barwedge L'.$$

We write

$$f : L \sim L'$$

to indicate that $f: L \to L'$ is a projectivity. If projectivity $f: L \sim L'$ carries a to a', b to b', c to c', . . . , we write

$$f : L(abc \cdots) \sim L'(a'b'c' \cdots).$$

Our first theorem tells us that projectivities can be found everywhere.

Theorem 1. *If a, b, c are three points on line L and a', b', c' are three points on L' in a projective plane, then there is a projectivity f such that*

$$f : L(abc) \sim L'(a'b'c').$$

PROOF. If $c = c'$ and $L \neq L'$, let $u = aa' \cap bb'$; then the perspectivity

$$L(abc) \overset{u}{\barwedge} L'(a'b'c')$$

Figure 2.2

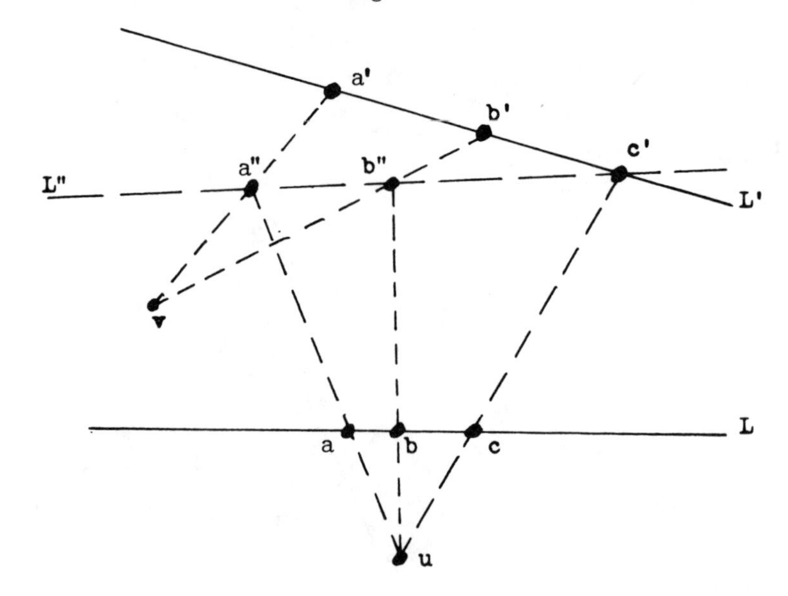

is such a projectivity. If c' is not on L, let L'' be a line on c' distinct from L' and not on point c (see Figure 2.2). Let u be a third point on cc', and let $a'' = ua \cap L''$, $b'' = ub \cap L''$, $v = a'a'' \cap b'b''$. Then

$$f : L(abc)\overset{u}{\barwedge}L''(a''b''c')\overset{v}{\barwedge}L'(a'b'c')$$

is a projectivity with the desired property. The cases c' on L with $c \neq c'$ and $L = L'$ are left as exercises. □

While Theorem 1 tells us that at least one projectivity exists, it does not say that only one such projectivity exists. The question of the uniqueness of the projectivity will be discussed in the next chapter.

For the time being, we shall content ourselves with a study of projectivities in the plane π_F over a field F. The first basic result is our next theorem.

Theorem 2. *Let L and L' be ranges in π_F. Let a and b be base points on L, and let a' and b' be base points on L'. If $f: L \sim L'$ is a projectivity, then there is a nonsingular 2×2 matrix M over F such that if c on L has parameters (λ, μ) relative to a and b and $f(c)$ on L' has parameters (λ', μ') relative to a' and b', then*

$$(\lambda', \mu') = (\lambda, \mu)M.$$

PROOF. If $f: L \sim L'$ is a projectivity, let

$$f : L \barwedge L_1 \barwedge L_2 \barwedge \cdots \barwedge L_n \barwedge L'$$

be a sequence of perspectivities whose composition is f. By Theorems 2.1.2 and 2.1.3, the perspectivity $L \barwedge L_1$ has an equation of the form

$$(\lambda_1, \mu_1) = (\lambda, \mu)M_1$$

with M_1 a 2×2 matrix over F. By Exercise 2.1.2, the matrix M_1 is nonsingular. Similarly, each perspectivity in the sequence has such an equation:

$$L_1 \barwedge L_2 : (\lambda_2, \mu_2) = (\lambda_1, \mu_1)M_2,$$
$$L_2 \barwedge L_3 : (\lambda_3, \mu_3) = (\lambda_2, \mu_2)M_3,$$
$$L_{n-1} \barwedge L_n : (\lambda_n, \mu_n) = (\lambda_{n-1}, \mu_{n-1})M_n,$$
$$L_n \barwedge L' : (\lambda', \mu') = (\lambda_n, \mu_n)M_{n+1}.$$

Telescoping these equations, we get

$$(\lambda', \mu') = (\lambda, \mu)M_1M_2 \cdots M_nM_{n+1}.$$

If we set

$$M = M_1M_2 \cdots M_nM_{n+1},$$

we have the desired equation. Since each M_i is nonsingular, M is nonsingular. □

The equation in Theorem 2 is called the *equation of the projectivity f*. As was true for perspectivities, the equation of f depends on the parametrizations of L and L'.

By applying Theorem 2.1.3 again, we note that the equation

$$(\lambda', \mu') = (\lambda, \mu)\begin{pmatrix} a & b \\ c & d \end{pmatrix}$$

can be written alternatively as

$$-b\lambda\lambda' + a\lambda\mu' - d\mu\lambda' + c\mu\mu' = 0.$$

Example 1. Find the equation of the projectivity that sends the points with parameters $(1,1)$, $(1,2)$, and $(0,1)$ to the points with parameters $(1,0)$, $(1,-1)$, and $(1,1)$, respectively.

SOLUTION. The projectivity must have an equation of the form

$$a\lambda\lambda' + b\lambda\mu' + c\mu\lambda' + d\mu\mu' = 0.$$

Requiring the corresponding pairs of parameters to satisfy this equation, we get the three conditions

$$\begin{aligned} a \quad\ + c \quad\quad &= 0, \\ a - b + 2c - 2d &= 0, \\ c + d &= 0. \end{aligned}$$

From the first and third we get

$$a = d = -c,$$

and from the second we get

$$-c - b + 2c + 2c = 0,$$

or

$$b = 3c.$$

Hence the equation is

$$-c\lambda\lambda' + 3c\lambda\mu' + c\mu\lambda' - c\mu\mu' = 0,$$

or

$$-\lambda\lambda' + 3\lambda\mu' + \mu\lambda' - \mu\mu' = 0.$$

The equivalent matrix form is

$$(\lambda', \mu') = (\lambda, \mu)\begin{pmatrix} 3 & 1 \\ -1 & -1 \end{pmatrix}.$$

ALTERNATIVE SOLUTION. The projectivity must have an equation of the form

$$(\lambda', \mu') = (\lambda, \mu)\begin{pmatrix} a & b \\ c & d \end{pmatrix}.$$

That means that

$$(1,1)\begin{pmatrix} a & b \\ c & d \end{pmatrix}$$

must be proportional to the parameters

$$(1,0),$$

or

$$(1,1)\begin{pmatrix} a & b \\ c & d \end{pmatrix} = k_1(1,0).$$

Similarly,

$$(1,2)\begin{pmatrix} a & b \\ c & d \end{pmatrix} = k_2(1,-1)$$

and

$$(0,1)\begin{pmatrix} a & b \\ c & d \end{pmatrix} = k_3(1,1).$$

These three matrix equations become the six linear equations

$$\begin{aligned}
a \qquad\quad + c \qquad\quad &= k_1,\\
b \qquad\quad + d &= 0,\\
a \qquad\quad + 2c \qquad\quad &= k_2,\\
b \qquad\quad + 2d &= -k_2,\\
c \qquad\quad &= k_3,\\
d &= k_3
\end{aligned}$$

in the unknowns a, b, c, d, k_1, k_2, and k_3. We need only solve for a, b, c, and d. From the last two equations, we have

$$c = d,$$

and from the second equation we have

$$b = -d.$$

Then the third and fourth equations yield

$$a + 2c = -(b + 2d),$$

or

$$\begin{aligned}
a &= -b - 2c - 2d\\
&= d - 2d - 2d\\
&= -3d.
\end{aligned}$$

Hence the matrix in the equation of the projectivity is

$$\begin{pmatrix} a & b \\ c & d \end{pmatrix} = \begin{pmatrix} -3d & -d \\ d & d \end{pmatrix}.$$

We may let d be any nonzero constant, since only proportionality counts. With $d = -1$, we have the same solution as before. □

We next ask the question whether any function defined by an equation of the form

$$(\lambda', \mu') = (\lambda, \mu)M$$

defines a projectivity. The answer is yes.

Theorem 3. *Let L and L' be ranges in π_F. Let a and b be base points on L, and let a' and b' be base points on L'. If M is a nonsingular 2×2 matrix over F, then the function $f: L \to L'$ defined by*

$$f(\lambda, \mu) = (\lambda, \mu)M$$

is a projectivity.

PROOF. First, we consider the case in which $L \neq L'$. One of the points a and b is not on L', say a. Let

$$M = \begin{pmatrix} x & y \\ z & w \end{pmatrix}$$

be given. The mapping f carries a to the point

$$f(a) = f(1,0) = (x, y).$$

Let L_1 be a line on point a distinct from L and not on $f(a)$ (see Figure 2.3). Let $L_1 \cap L'$ be the point (s,t), in terms of the base points a' and b'. Then $(s,t) \neq (x, y)$, so

$$sy - tx \neq 0.$$

The plan of our proof is now to construct two perspectivities, $f_1 : L \overline{\wedge} L_1$ with equation $f_1(\lambda, \mu) = (\lambda, \mu)M_1$, and $f_2 : L_1 \overline{\wedge} L'$ with equation $f_2(\lambda_1, \mu_1) = (\lambda_1, \mu_1)M_2$, for appropriately chosen matrices M_1 and M_2, such that $f = f_2 \cdot f_1$, or equivalently, $M_1 M_2$ is proportional to M. To this end, let line L_1 have base points a and (s,t), and define f_1 by the matrix

$$M_1 = \begin{pmatrix} sy - tx & 0 \\ sw - tz & -1 \end{pmatrix}.$$

Note that

$$\det(M_1) = (sy - tx)(-1) \neq 0,$$

Figure 2.3

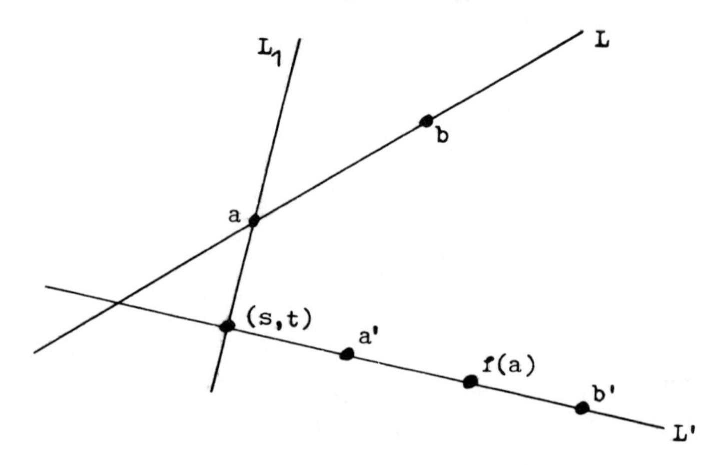

so M_1 is nonsingular. The equation

$$p_1\alpha - p_2\beta + p_3\gamma - p_4\delta = 0$$

of Theorem 2.1.4 is also satisfied, for $\alpha = 0$ and, a being the first base point on both L and L', $p_2 = p_3 = p_4 = 0$. Hence f_1 is indeed a perspectivity. Using the same base points on L_1 as before, we now define f_2 by the matrix

$$M_2 = \begin{pmatrix} x & y \\ s(xw - yz) & t(xw - yz) \end{pmatrix}.$$

Note that

$$\begin{aligned} \det(M_2) &= xt(xw - yz) - sy(xw - yz) \\ &= (xt - sy)\begin{vmatrix} x & y \\ z & w \end{vmatrix} \neq 0, \end{aligned}$$

so M_2 is nonsingular. Since the point (s, t) has coordinates

$$[sa_1' + tb_1', sa_2' + tb_2', sa_3' + tb_3'],$$

the second row of the determinant defining p_1 in Theorem 2.1.4 has elements of the form

$$\begin{vmatrix} sa_i' + tb_i' & sa_j' + tb_j' \\ a_i' & a_j' \end{vmatrix} = t\begin{vmatrix} b_i' & b_j' \\ a_i' & a_j' \end{vmatrix},$$

and the third row of the same determinant has elements of the form

$$\begin{vmatrix} sa_i' + tb_i' & sa_j' + tb_j' \\ b_i & b_j \end{vmatrix} = s\begin{vmatrix} a_i' & a_j' \\ b_i' & b_j' \end{vmatrix}.$$

Hence the last two rows of the determinant defining p_1 are linearly dependent (s times the second plus t times the third equals zero), so $p_1 = 0$. Similarly, $p_2 = 0$. The third row of the determinant defining p_3 has elements of the form

$$\begin{vmatrix} sa_i' + tb_i' & sa_j' + tb_j' \\ b_i' & b_j' \end{vmatrix} = s\begin{vmatrix} a_i' & a_j' \\ b_i' & b_j' \end{vmatrix},$$

and the third row of the determinant defining p_4 has elements of the form

$$\begin{vmatrix} sa_i' + tb_i' & sa_j' + tb_j' \\ a_i' & a_j' \end{vmatrix} = t\begin{vmatrix} b_i' & b_j' \\ a_i' & a_j' \end{vmatrix}.$$

Hence $tp_3 = -sp_4$, or

$$tp_3 + sp_4 = 0.$$

Hence, for f_2,

$$\begin{aligned} p_1\alpha - p_2\beta + p_3\gamma - p_4\delta &= p_3[-t(xw - yz)] - p_4[s(xw - yz)] \\ &= -(tp_3 + sp_4)(xw - yz) \\ &= 0, \end{aligned}$$

and f_2 is a perspectivity by Theorem 2.1.4. Finally, we have

$$\begin{aligned} M_1M_2 &= \begin{pmatrix} sy-tx & 0 \\ sw-tz & -1 \end{pmatrix}\begin{pmatrix} x & y \\ s(xw-yz) & t(xw-yz) \end{pmatrix} \\ &= \begin{pmatrix} (sy-tx)x & (sy-tx)y \\ (sw-tz)x-s(xw-yz) & (sw-tz)y-t(xw-yz) \end{pmatrix} \\ &= \begin{pmatrix} (sy-tx)x & (sy-tx)y \\ (sy-tx)z & (sy-tx)w \end{pmatrix} \\ &= (sy-tx)M, \end{aligned}$$

so $f = f_2 \cdot f_1$ and f is a projectivity. This takes care of the case $L \neq L'$. If $L = L'$, let L_2 be any line distinct from L, and let $g: L \barwedge L_2$ be any perspectivity, say with matrix N. Then

$$f \cdot g^{-1}: L_2 \to L'$$

is a function defined by

$$f \cdot g^{-1}(\lambda_2, \mu_2) = (\lambda_2, \mu_2)N^{-1}M,$$

so by what we have proved already it is the composition of two perspectivities,

$$f \cdot g^{-1} = f_2 \cdot f_1.$$

Hence

$$f = f_2 \cdot f_1 \cdot g,$$

a composition of three perspectivities, so f is a projectivity. □

Exercises 2.2

***1.** Show that a projectivity from L to L' is a bijection and that its inverse is a projectivity from L' to L.

2. Prove Theorem 1 for the case c' on L and $c \neq c'$.

3. Carry out a proof of Theorem 1 for the case $L = L'$.

4. Let line L in π_R have base points $a = [-1, 1, 1]$ and $b = [1, 0, 0]$, and let line L' have base points $a' = [2, 1, 0]$ and $b' = [2, 1, 1]$. If projectivity $f: L \sim L'$ has equation

$$(\lambda', \mu') = (\lambda, \mu)\begin{pmatrix} 4 & -2 \\ 0 & -5 \end{pmatrix},$$

find the images of points a, b, $[1, 1, 1]$, and $[0, 1, 1]$.

5. Let L, L', a, b, a', and b' be as in Problem 4. If projectivity $f: L \sim L'$ carries $[1, 1, 1]$ to $[2, 1, 0]$, $[1, 0, 0]$ to $[0, 0, 1]$, and $[0, 1, 1]$ to $[2, 1, -1]$, find the equation of f.

***6.** Show that the projectivity in Theorem 1 is unique in the plane π_F.

*7. Let $f: L \sim L$ be a projectivity from a line L to itself. We call f an *involution* in case f is not the identity mapping but $f^2 = f \cdot f$ is the identity. Show that if f is an involution and $f(p) = q$, then $f(q) = p$.

8. Prove that the projectivity $f: L \to L$ with equation

$$f(\lambda, \mu) = (\lambda, \mu)\begin{pmatrix} a & b \\ c & d \end{pmatrix}$$

is an involution if and only if $a + d = 0$.

9. Show that

$$(\lambda, \mu)\begin{pmatrix} a & b \\ c & d \end{pmatrix} = k(\lambda, \mu)$$

has a solution for λ and μ only for those values of k satisfying

$$\begin{vmatrix} a-k & b \\ c & d-k \end{vmatrix} = 0.$$

10. If $f: L \sim L$ is a projectivity from a line to itself, a *fixed point* of f is a point p on on L such that $f(p) = p$. If f has the equation

$$f(\lambda, \mu) = (\lambda, \mu)\begin{pmatrix} 2 & -1 \\ -1 & 2 \end{pmatrix},$$

find the parameters of all the fixed points of f.

11. Repeat Problem 9 for the equation

$$f(\lambda, \mu) = (\lambda, \mu)\begin{pmatrix} 4 & -3 \\ 0 & -4 \end{pmatrix}.$$

12. If L is a line in π_F, how many fixed points may a projectivity $f: L \sim L$ have?

13. Show that if $f: L \sim L$ is an involution in π_R with a fixed point p, then f has a second fixed point distinct from p.

Section 2.3. Collineations

We are now ready to consider functions from a projective plane to itself. Let π be a projective plane. A *collineation* on π is an isomorphism from π to itself. We shall denote a collineation (f, F) on π as

$$f: \pi \to \pi,$$

recalling from Theorem 1.4.1 that f is a bijection that preserves collinearity.

We say a collineation $f: \pi \to \pi$ *fixes point* x in case $f(x) = x$. We say f *fixes line* L in case, for every point x on L, $f(x)$ is also on L [equivalently, $F(L) = L$]. We say line L is *pointwise fixed* by f in case, for every point x on L, f fixes x. We say point x is *linewise fixed* by f in case, for every line L on x, f fixes L.

Theorem 1. *Let $f: \pi \to \pi$ be a collineation.*

a. *If f fixes two points p and q, then f fixes line pq.*
b. *If f fixes two lines L and M, then f fixes point $L \cap M$.*
c. *If two lines are both pointwise fixed by f, then f is the identity function.*
d. *If two points are both linewise fixed by f, then f is the identity function.*

PROOF.

a. Let x be any point on pq. Since p, q, and x are collinear and f preserves collinearity, $f(p)$, $f(q)$, and $f(x)$ are collinear. That is, since $f(p) = p$ and $f(q) = q$, $f(x)$ is on pq. Thus f fixes pq.

b. Let $p = L \cap M$. Since p is on L and f fixes L, $f(p)$ is on L. Similarly $f(p)$ is on M. Hence $f(p) = L \cap M = p$, and p is fixed.

c. To show that f is the identity, it will be sufficient to show that f fixes every point. Let L and M be the two pointwise fixed lines, and let x be any point. If x is on L or M, x is fixed. If x is not on L or M, let L_1 and L_2 be two lines on x. If $p = L_1 \cap L$ and $q = L_1 \cap M$, then p and q are fixed, so by part a, L_1 is fixed. Similarly, L_2 is fixed. Hence by part b, x is fixed.

d. Again, we will show that f fixes every point. Let p and q be the two linewise fixed points. Then p and q are fixed. (Why?) If x is a point not on line pq, then the two lines px and qx are fixed, so by part a, x is fixed. If x is a third point on line pq, let L be any line on x distinct from pq. Since every point on L, with the possible exception of x, is fixed by what we have just proved, L is fixed. Since pq is fixed, we again have by part a that x is fixed. □

We ought to point out that a line can be fixed by a collineation without being pointwise fixed. That is, the points on the fixed line may be shuffled around, but have images on the same line, without each point being its own image.

One question that arises is whether any of the points on a fixed line must be fixed points. We shall not be able to completely answer that question in this chapter, but the following results are a step in that direction, unrelated as they may seem at this point.

Theorem 2. *If the collineation $f: \pi \to \pi$ holds some line pointwise fixed, then f also holds some point linewise fixed.*

PROOF. Suppose f holds line A pointwise fixed. Either some point not on A is fixed or no point on A is fixed by f. If c is a point not on A that is fixed by f, then c is linewise fixed. For, if L is any line on c, let $L \cap A = p$. Since p and c are fixed, L is fixed. On the other hand, if no point not on A is fixed by f, let p be any point not on A, and let $p' = f(p)$. Then $p' \neq p$. Let line pp' meet A at c (see Figure 2.4). We first note that line pp' is fixed. For if x is any point on pp', then x, p, and c are collinear. Hence $f(x)$, $f(p) = p'$, and $f(c) = c$ are collinear, so that $f(x)$ is on $p'c = pp'$. We now wish to show that c is held linewise fixed by f. Let L be any line on c, other than A or pp'. If x is any point on L, suppose $x' = f(x)$ is not on L. Then line xx' is fixed, as we argued for pp', and the point $q = xx' \cap pp'$ is fixed. But q is not on A, and we are assuming that only points on A are fixed by f. Hence x' must be on L, and L is fixed. We have thus shown that c is linewise fixed. □

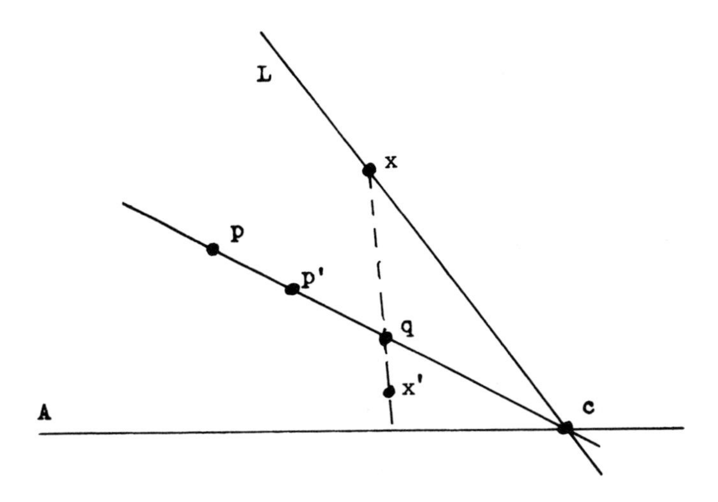

Figure 2.4

Theorem 3. *Let $f : \pi \to \pi$ be a collineation, and let L be a range in π. If the restriction $f|_L$ of f to L is a projectivity, then for any range M in π, $f|_M$ is also a projectivity.*

PROOF. Given that $f|_L$ is a projectivity, let M be any range in π. Let c be a point not on L or M, and let g be the perspectivity $g : L \overset{c}{\barwedge} M$. Let $L' = f(L)$, $M' = f(M)$, and $c' = f(c)$ (see Figure 2.5). Let $h = f|_M \cdot g \cdot f|_L^{-1} : L' \to L \to M \to M'$. If p is a point on L, and $g(p) = q$ on M, then line pq is on c. If $p' = f(p)$ and $q' = f(q)$, then p', q', and c' are collinear. Hence h is the perspectivity $h : L' \overset{c'}{\barwedge} M'$, for $h(p') = f|_M \cdot g \cdot f|_L^{-1}(p') = f|_M g(p) = f|_M(q) = q'$. From $h = f|_M \cdot g \cdot f|_L^{-1}$, we get $f|_M = h \cdot f|_L \cdot g^{-1}$. Now h is a perspectivity, $f|_L$ is a projectivity, and g^{-1} is a perspectivity, so $f|_M$ is a projectivity. □

A collineation $f : \pi \to \pi$ for which $f|_L$ is a projectivity for some range L (and hence for all ranges) in π is called a *projective* collineation. For example, a collineation that holds some line pointwise fixed is a projective

Figure 2.5

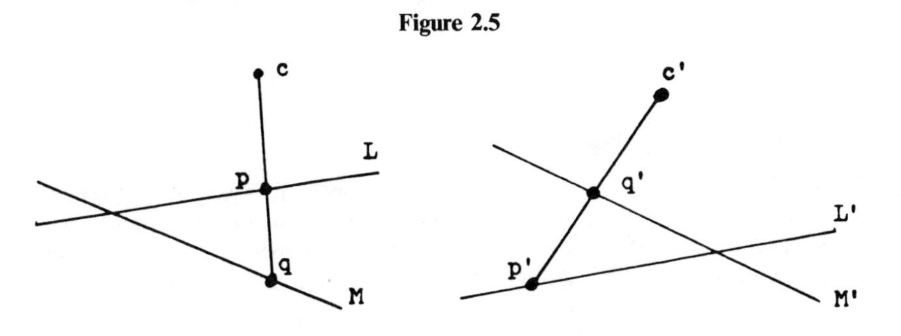

collineation, for its restriction to the pointwise fixed line is the identity function, which is a projectivity.

Exercises 2.3

*1. Verify that the identity function from a line to itself is a projectivity.

*2. Verify that the identity function from a projective plane to itself is a collineation.

3. Prove the converse of Theorem 2.

*4. Prove that the collineations on the projective plane π form a group under function composition. [We denote this group by $C(\pi)$.]

5. Is the set of projective collineations on π a group?

Section 2.4. Matrix-Induced Collineations on π_F.

In this section and the next three, we shall consider some particular kinds of collineations. In this section, we shall discuss collineations on the plane π_F that are defined algebraically by the use of matrices. Let

$$M = \begin{bmatrix} a_{11} & a_{12} & a_{13} \\ a_{21} & a_{22} & a_{23} \\ a_{31} & a_{32} & a_{33} \end{bmatrix}$$

be a 3×3 matrix over F, and let (x_1, x_2, x_3) be a vector of length 3 with entries in F. By $(x_1, x_2, x_3)M$ we mean the matrix product

$$\begin{aligned}(x_1, x_2, x_3)\begin{bmatrix} a_{11} & a_{12} & a_{13} \\ a_{21} & a_{22} & a_{23} \\ a_{31} & a_{32} & a_{33} \end{bmatrix} &= (a_{11}x_1 + a_{21}x_2 + a_{31}x_3, a_{12}x_1 + a_{22}x_2 \\ &\quad + a_{32}x_3, a_{13}x_1 + a_{23}x_2 + a_{33}x_3).\end{aligned}$$

If $x = (x_1, x_2, x_3)$ is any 3-vector over F, then by $[x] = [(x_1, x_2, x_3)]$ we mean the point $[x_1, x_2, x_3]$ in the projective plane π_F.

We are now ready to define a matrix-induced collineation. Let M be a 3×3 nonsingular matrix over F. Let $f_M : \pi_F \to \pi_F$ be defined by

$$f_M[x_1, x_2, x_3] = [(x_1, x_2, x_3)M].$$

Theorem 1. *f_M is a collineation.*

PROOF. We first show that f_M is one-to-one. Suppose $f_M[x_1, x_2, x_3] = f_M[y_1, y_2, y_3]$. Then $[(x_1, x_2, x_3)M] = [(y_1, y_2, y_3)M]$, so there is a constant of proportionality k such that $(x_1, x_2, x_3)M = k(y_1, y_2, y_3)M$. Since M is nonsingular, we can multiply on the right by M^{-1}, getting $(x_1, x_2, x_3) = k(y_1, y_2, y_3)$. Hence $[x_1, x_2, x_3]$

$=[y_1, y_2, y_3]$, and f_M is one-to-one. Next we show that f_M is onto. If $[y_1, y_2, y_3]$ is any point, then also $[(y_1, y_2, y_3)M^{-1}]$ is a point and $f_M[(y_1, y_2, y_3)M^{-1}]=[y_1, y_2, y_3]$. Hence f is onto. Thus f_M is a bijection. Finally, we show that f_M preserves collinearity. Let $[x_{i1}, x_{i2}, x_{i3}]$ $(i=1,2,3)$ be three collinear points. Then their images $[(x_{i1}, x_{i2}, x_{i3})M]$ $(i=1,2,3)$ are collinear, for the determinant

$$\begin{vmatrix} (x_{11}, x_{12}, x_{13})M \\ (x_{21}, x_{22}, x_{23})M \\ (x_{31}, x_{32}, x_{33})M \end{vmatrix} = \begin{vmatrix} x_{11} & x_{12} & x_{13} \\ x_{21} & x_{22} & x_{23} \\ x_{31} & x_{32} & x_{33} \end{vmatrix} |M|$$

is zero because the determinant

$$\begin{vmatrix} x_{11} & x_{12} & x_{13} \\ x_{21} & x_{22} & x_{23} \\ x_{31} & x_{32} & x_{33} \end{vmatrix}$$

is zero. Here we have used Theorem 1.3.3 and the fact that the determinant of a product of square matrices is the product of their determinants. □

It is f_M that is called a *matrix-induced collineation*. Some of the properties of matrix-induced collineations are given by the following theorems.

Theorem 2. *If M and N are nonsingular 3×3 matrices over F, then the matrix-induced collineations f_M and f_N are identical if and only if M is a scalar multiple of N, or there exists $k \in F$ such that $M = kN$.*

PROOF.

$$\begin{aligned} f_M = f_N &\Leftrightarrow f_M[x_1, x_2, x_3] = f_N[x_1, x_2, x_3] \\ &\Leftrightarrow [(x_1, x_2, x_3)M] = [(x_1, x_2, x_3)N] \\ &\Leftrightarrow (x_1, x_2, x_3)M = k(x_1, x_2, x_3)N \\ &\Leftrightarrow (x_1, x_2, x_3)(M-kN) = 0, \end{aligned}$$

for some $k \in F$ and for all points $[x_1, x_2, x_3]$. If $M = kN$, then $M - kN = 0$ and $f_M = f_N$. Conversely, if $f_M = f_N$, then $(x_1, x_2, x_3)(M-kN)=0$ for the triples $(x_1, x_2, x_3) = (1,0,0)$, $(0,1,0)$, and $(0,0,1)$. Hence each row of $M-kN$ is zero, so that $M-kN$ is the zero matrix. Hence $M = kN$. □

Theorem 3. *A matrix-induced collineation is projective.*

PROOF. Let

$$M = \begin{bmatrix} a & b & c \\ d & e & f \\ g & h & i \end{bmatrix}.$$

Let L be the line $\langle 0,0,1\rangle$, and let $[1,0,0]$ and $[0,1,0]$ be base points on L. Let $[a,b,c]$ and $[d,e,f]$ be base points on $L' = f_M(L)$. Then the point on L with parameters (λ, μ) is the point $[\lambda, \mu, 0]$ and has image

$$\begin{aligned} f_M[\lambda, \mu, 0] &= [\lambda a + \mu d, \lambda b + \mu e, \lambda c + \mu f] \\ &= [\lambda(a,b,c) + \mu(d,e,f)]. \end{aligned}$$

Hence $f_M[\lambda, \mu, 0]$ has parameters $(\lambda', \mu') = (\lambda, \mu)$ relative to $[a, b, c]$ and $[d, e, f]$. Thus $f_M|_L$ has the equation

$$(\lambda', \mu') = (\lambda, \mu)\begin{pmatrix} 1 & 0 \\ 0 & 1 \end{pmatrix},$$

so by Theorem 2.2.3 is a projectivity. Hence f_M is projective. □

We next consider the question of fixed points of a matrix-induced collineation f_M. Suppose that f_M fixes points $[x_1, x_2, x_3]$. Then

$$f_M[x_1, x_2, x_3] = [x_1, x_2, x_3],$$

so that

$$[(x_1, x_2, x_3)M] = [x_1, x_2, x_3].$$

Hence for some constant k,

$$(x_1, x_2, x_3)M = k(x_1, x_2, x_3),$$

or

$$(x_1, x_2, x_3)(M - kI) = 0, \tag{1}$$

I being the 3×3 identity matrix.

We expand the matrix equation (1) in order to scrutinize it more closely. If

$$M = \begin{bmatrix} a_{11} & a_{12} & a_{13} \\ a_{21} & a_{22} & a_{23} \\ a_{31} & a_{32} & a_{33} \end{bmatrix},$$

then

$$M - kI = \begin{bmatrix} a_{11} - k & a_{12} & a_{13} \\ a_{21} & a_{22} - k & a_{23} \\ a_{31} & a_{32} & a_{33} - k \end{bmatrix},$$

and Equation (1) becomes the system of equations

$$\begin{aligned} (a_{11} - k)x_1 + {} & a_{21}x_2 + a_{31}x_3 = 0, \\ a_{12}x_1 + {} & (a_{22} - k)x_2 + a_{32}x_3 = 0, \\ a_{13}x_1 + {} & a_{23}x_2 + (a_{33} - k)x_3 = 0. \end{aligned} \tag{2}$$

The system (2) has a nontrivial solution if and only if the determinant of coefficients is zero, i.e.,

$$\begin{vmatrix} a_{11} - k & a_{21} & a_{31} \\ a_{12} & a_{22} - k & a_{32} \\ a_{13} & a_{23} & a_{33} - k \end{vmatrix} = |M - kI| = 0. \tag{3}$$

This last condition is called the *characteristic equation* of the matrix M; it is a cubic equation in k. The values of k that satisfy (3) are called the *eigenvalues* of M. For each eigenvalue k, any corresponding solution (x_1, x_2, x_3) of the system (2) is called an *eigenvector* of M. Thus if $[x_1, x_2, x_3]$ is a fixed point of f_M, then (x_1, x_2, x_3) must be an eigenvector of M. Since the converse of this statement is also true, we have the following theorem.

Theorem 4. *$[x_1, x_2, x_3]$ is a fixed point of f_M if and only if (x_1, x_2, x_3) is an eigenvector of M.*

The problem of finding the fixed points of f_M therefore reduces to the problem of finding eigenvectors of M. Consider an example.

Example 1. Let

$$M = \begin{bmatrix} 2 & 1 & 6 \\ 1 & -1 & 1 \\ -1 & 2 & 0 \end{bmatrix}$$

induce the collineation f_M on π_R. To find the fixed points of f_M, we first find the eigenvalues of M from equation (3):

$$\begin{vmatrix} 2-k & 1 & 6 \\ 1 & -1-k & 1 \\ -1 & 2 & -k \end{vmatrix} = -k^3 + k^2 - k + 1$$

$$= (k^2+1)(-k+1) = 0.$$

The only real eigenvalue is thus $k = 1$. The corresponding eigenvectors, from the system (2), are found by solving

$$\begin{aligned} x_1 + x_2 - x_3 &= 0, \\ x_1 - 2x_2 + 2x_3 &= 0, \\ 6x_1 + x_2 - x_3 &= 0. \end{aligned}$$

Thus, any eigenvector is of the form

$$(0, a, a),$$

which corresponds to the point

$$[0, 1, 1].$$

This is indeed a fixed point of f_M, as you may verify directly. It is the only fixed point.

We should note that if f_M in Example 1 had been defined on π_C instead of π_R, there would have been two more fixed points, corresponding to the complex eigenvalues i and $-i$.

Exercises 2.4

1. Show that $f_N \cdot f_M = f_{MN}$.

2. Find all the fixed points of $f_M : \pi_R \to \pi_R$, given

$$\textbf{(a)}\ M = \begin{pmatrix} 1 & 3 & -2 \\ 0 & 1 & -1 \\ 0 & 0 & 2 \end{pmatrix}, \qquad \textbf{(b)}\ M = \begin{pmatrix} 4 & 3 & -1 \\ -4 & -4 & 2 \\ -6 & -9 & 5 \end{pmatrix},$$

$$\textbf{(c)}\ M = \begin{pmatrix} 1 & -2 & 2 \\ 1 & -2 & 1 \\ -2 & 2 & -3 \end{pmatrix}.$$

***3.** Show that line $L = \langle l_1, l_2, l_3 \rangle$ is fixed by f_M if and only if

$$M^{-1} \begin{bmatrix} l_1 \\ l_2 \\ l_3 \end{bmatrix} = k \begin{bmatrix} l_1 \\ l_2 \\ l_3 \end{bmatrix}.$$

***4.** If

$$M = \begin{pmatrix} 1 & 0 & 0 \\ 0 & 1 & 0 \\ r & s & t \end{pmatrix}$$

is a nonsingular matrix, show that f_M holds line $\langle 0, 0, 1 \rangle$ pointwise fixed and point $[r, s, t-1]$ linewise fixed.

5. If

$$M = \begin{bmatrix} h_1x_1 & h_1x_2 & h_1x_3 \\ h_2y_1 & h_2y_2 & h_2y_3 \\ h_3z_1 & h_3z_2 & h_3z_3 \end{bmatrix},$$

show that f_M carries

$$\begin{aligned} &[1,0,0] \text{ to } [x_1, x_2, x_3], \\ &[0,1,0] \text{ to } [y_1, y_2, y_3], \\ &[0,0,1] \text{ to } [z_1, z_2, z_3]. \end{aligned}$$

***6.** In Exercise 5, set up a system of equations in h_1, h_2, h_3 whose solution, when substituted into the matrix M, makes M such that f_M carries $[1, 1, 1]$ to $[p_1, p_2, p_3]$. What restrictions are there on point $[p_1, p_2, p_3]$?

7. Construct a matrix M so that f_M carries:

(a) $[1,0,0]$ to $[2,-1,0]$, $[0,1,0]$ to $[3,1,1]$, $[0,0,1]$ to $[0,1,-1]$, and $[1,1,1]$ to $[2,1,-1]$.

(b) $[1,1,0]$ to $[0,1,-1]$, $[0,1,2]$ to $[3,2,0]$, $[1,0,1]$ to $[2,2,1]$, and $[1,1,-1]$ to $[1,-1,1]$.

***8.** Prove that there is always a matrix-induced collineation on π_F that carries a given four-point onto a given image four-point.

Section 2.5. Central Collineations

A collineation that holds some line pointwise fixed (and consequently holds some point linewise fixed) is called a *central collineation*. The pointwise fixed line is called the *axis* and the linewise fixed point is called the *center* of the central collineation.

If the center of a central collineation lies on the axis, we call the collineation an *elation*; if it does not, we call the collineation a *homology*. (The identity function is regarded as both an elation and a homology.) If $f:\pi\to\pi$ is a central collineation that is not the identity, such that $f^2=f\cdot f$ is the identity, we call f *harmonic*.

The basic properties of central collineations are contained in the following theorems and in the exercises.

Theorem 1. *Let $f:\pi\to\pi$ be a central collineation with center c and axis A. If x is any point of π, then $f(x)$ is on xc; if L is any line in π, then $f(L)$ is on $L\cap A$.*

PROOF. Since c is linewise fixed, xc is fixed by f, and $f(x)$ is on xc. Since A is pointwise fixed, $L\cap A$ is fixed by f. Since $L\cap A$ is on L, $L\cap A$ is also on $f(L)$, or $f(L)$ is on $L\cap A$. □

Theorem 2. *A central collineation is projective.*

PROOF. If f is a central collineation with axis A, then $f|_A$ is the identity, which is a projectivity. □

Exercises 2.5

1. Prove that a central collineation that is not the identity has a unique axis and a unique center.

2. Given the axis A of an elation f and one point p and its image $p'=f(p)$, construct the center c of f and the image $x'=f(x)$ of an arbitrary point x.

3. Given the axis A and center c of a homology f and one point p and its image $p'=f(p)$, construct the image $x'=f(x)$ of an arbitrary point x.

4. Given four points p, q, p', and q', does there exist a central collineation f such that $f(p)=p'$ and $f(q)=q'$?

5. Given six points p, q, r, p', q', and r', such that pp', qq', and rr' are concurrent, does there exist a central collineation f such that $f(p)=p'$, $f(q)=q'$, and $f(r)=r'$?

6. Given point $c=[c_1,c_2,c_3]$ and line $A=\langle a_1,a_2,a_3\rangle$ in π_F, construct a matrix M such that $f_M:\pi_F\to\pi_F$ will be a central collineation with center c and axis A.

7. Is the set of central collineations on the projective plane π a group under function composition?

Section 2.6. Central Collineations on π_R

In this section we shall consider some central collineations on π_R. They are of interest because of their action on α_R, the Euclidean plane, which is embedded in π_R.

A *plane perspective* is a central collineation on π_R whose center and axis are both finite. The line whose image under a plane perspective is the ideal line is called the *vanishing line*.

The first problem we shall tackle is that of constructing the vanishing line of a plane perspective. First note that if V is the vanishing line of a plane perspective f with axis A, then by Theorem 2.5.1, $f(V)$ is on $V \cap A$. But $f(V)$ is the ideal line, so point $V \cap A$ must be an ideal point. That is, V is parallel to A, from the Euclidean point of view.

Thus given the axis A and center c of a plane perspective f, together with a point p and its image $p' = f(p)$ (see Figures 2.6 and 2.7), we may construct the vanishing line V as follows: Let L be a line on c which meets A; then L also meets V, at some point v, yet to be determined. Choosing any likely point as v, we construct the image of v by letting $q = pv \cap A$, $v' = f(v) = p'q \cap L$. But v' must be an ideal point, so $p'q$ must be parallel

Figure 2.6

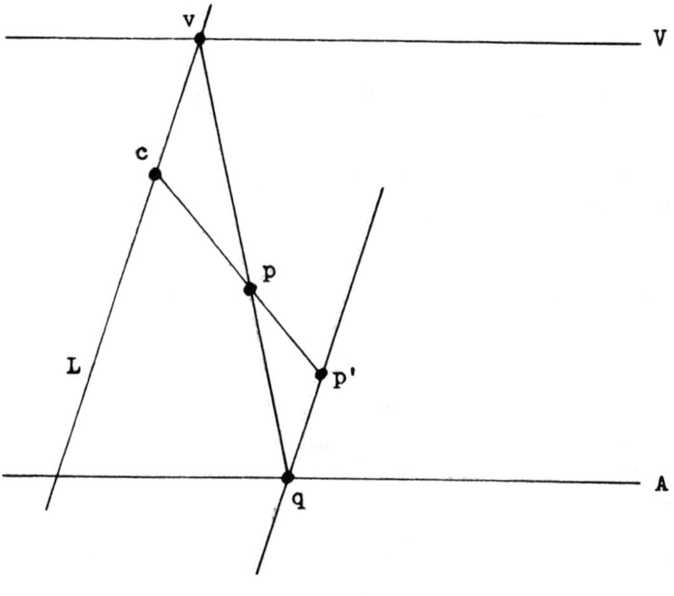

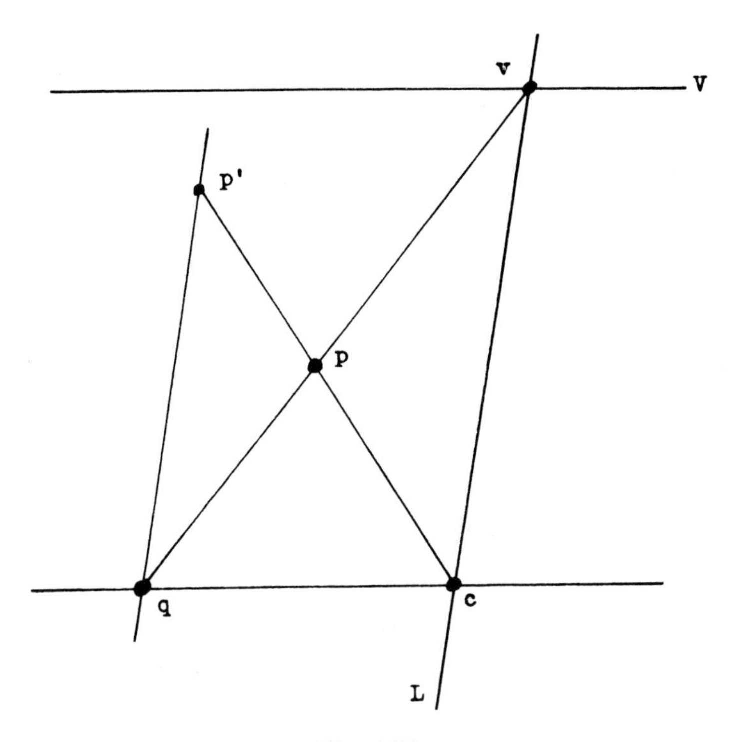

Figure 2.7

to L. Hence, working backward, let q be the point on A such that $p'q \| L$. Then $v = pq \cap L$. Hence V is the line on v, parallel to A.

Plane perspectives can be constructed that carry out certain transformations of figures in the Euclidean plane. For example, a pair of intersecting lines L and M can be transformed into parallel lines by requiring the vanishing line of a plane perspective to pass through $L \cap M$. That is, $f(L)$ and $f(M)$ will be parallel because the point $f(L) \cap f(M) = f(L \cap M)$ is ideal. The key to such constructions is the following theorem.

Theorem 1. *If f is a plane perspective with axis A, center c, and vanishing line V, suppose line L meets A at p and V at q. Then $f(L)$ is the line on p that is parallel to qc.*

PROOF. Since lines L and qc meet at q on the vanishing line, $f(L)$ and $f(qc)$ must be parallel because $f(q)$ is an ideal point. But any line on c is fixed, so $f(qc) = qc$. Thus $f(L)$ is parallel to qc. Also, $f(L)$ is on p by Theorem 2.5.1. □

Example 1. Suppose $pqrs$ is a quadrilateral in α_R that is not a trapezoid. We will construct a plane perspective that transforms $pqrs$ into a square. Let $pq \cap rs = v_1$, $qr \cap sp = v_2$ (see Figure 2.8). By letting $V = v_1v_2$ be the vanishing line, we automatically transform $pqrs$ into a parallelogram. In

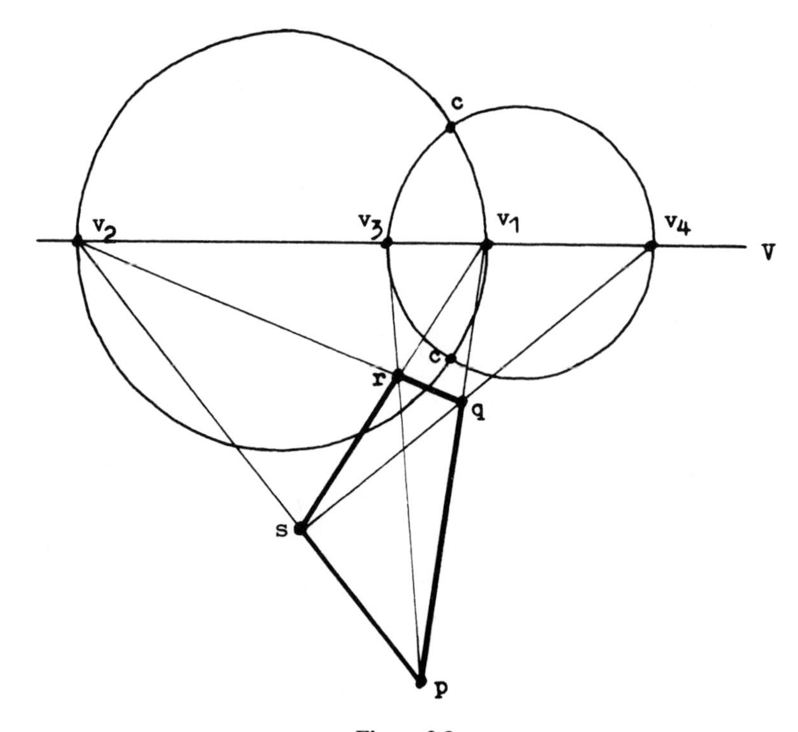

Figure 2.8

order to further require the image to be a rectangle, it is sufficient to give the image parallelogram one right angle. In order to do this, we recall Theorem 1; if c is the center of the plane perspective, then line pq has image parallel to cv_1 and line ps has image parallel to cv_2. Thus it suffices to select c so that $\angle v_1cv_2$ is a right angle. The locus of such points c is the circle with diameter the segment $\overline{v_1v_2}$. Thus selecting c on this circle will force the image to be a rectangle. Now to get the image to be a square, we note that of all rectangles, only the square has perpendicular diagonals. Hence, we let $v_3 = pr \cap V$ and $v_4 = qs \cap V$; to force pr and qs to have images that are perpendicular to each other, it suffices to select the center c so that $\angle v_3cv_4$ is a right angle. This places c on the circle with diameter $\overline{v_3v_4}$. Hence, to transform $pqrs$ into a square, we select c to be either of the intersection points of the two circles. Our choice of axis (or equivalently, image of p; see Exercises 1 and 2) is now almost arbitrary; the selection of V and c as described forces the image of $pqrs$ to be a square.

A *dilatation* is a central collineation on π_R whose axis is the ideal line. A dilatation is called a *homothecy* or a *translation* according as its center is finite or ideal.

Because the ideal line is fixed under a dilatation, we may regard a dilatation as a transformation from α_R to itself. It is in this context that we state some properties of dilatations.

Theorem 2. *Under a dilatation, the image of a line L is either L itself or a line parallel to L.*

PROOF. If the image of L is not L itself, then L and its image meet on the ideal line (Theorem 2.5.1). □

Theorem 3. *In α_R, let (xy) denote the directed distance from x to y. If f is a homothecy, then there is a constant r such that if p and q are finite points and $p' = f(p)$, $q' = f(q)$, then $(p'q') = r(pq)$.*

PROOF. Let f have center c, and suppose first that $pq \neq p'q'$. Then $p'q' \| pq$, so $\triangle pqc$ is similar to $\triangle p'q'c$ (see Figure 2.9). Let $r = (p'c)/(pc)$. Then $r = (p'q')/(pq)$, or $(p'q') = r(pq)$. The fact that r is actually independent of p is easily shown, using similar triangles again. Finally, if p, q, and c are collinear, let s be a point not on pc. Then the original conclusion can be reached, using two sets of similar triangles. This step will be left as an exercise. □

The number r in Theorem 3 is called the *homothetic ratio* of f. If $r > 0$, the homothecy f is called *direct*; if $r < 0$, f is called *indirect*. It can be shown that f is the identity if $r = 1$, and f is harmonic if and only if $r = -1$.

Other properties of central collineations on π_R are found in the exercises.

Exercises 2.6

1. Given the axis, center, and vanishing line of a plane perspective, construct the image of an arbitrary point.

Figure 2.9

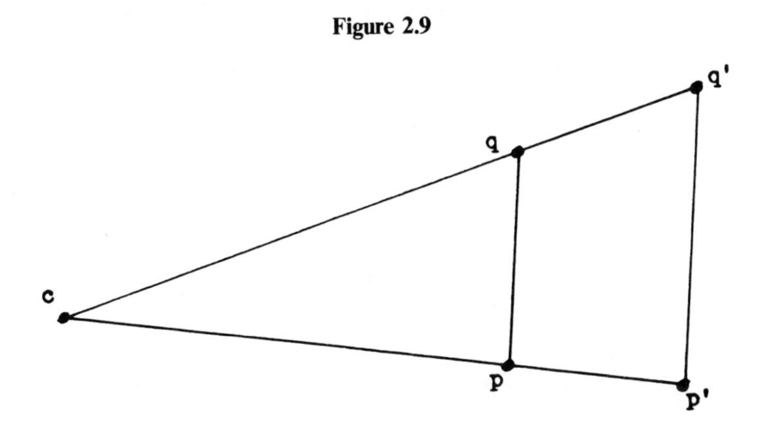

2. Given the axis and vanishing line of a plane perspective, together with one point and its image, construct the center.

3. Given the center and vanishing line of a plane perspective, together with one point and its image, construct the axis.

4. Construct a plane perspective that transforms a given triangle into an isosceles right triangle.

5. Construct a plane perspective that transforms a given triangle into an equilateral triangle.

6. Show that the image of a circle under a plane perspective is a parabola, hyperbola, or ellipse according as the circle meets the vanishing line in one, two, or no points. (See Exercise 1.7.6.)

7. In α_R, let line $x = 2$ be the axis, point $(0, 1)$ the center, and line $x = -1$ the vanishing line of a plane perspective. Find the image of an arbitrary point (a, b), and find the image locus of the circle $x^2 + y^2 = 1$.

8. A circle is transformed into a parabola by a plane perspective. Construct the line whose image is the axis of the image parabola.

9. A circle is transformed into a hyperbola by a plane perspective. Construct the points on the circle whose images are the vertices of the image hyperbola.

10. Prove that a dilatation preserves angles.

11. Prove that the image of a circle under a dilatation is a circle.

12. Prove that a translation preserves distances. Does this fact justify the use of the term "translation"?

13. In the proof of Theorem 3, verify that r is independent of point p.

14. Complete the proof of Theorem 3 by treating the case of p, q, c collinear.

15. Prove that a homothecy is harmonic if and only if its ratio is -1.

16. Prove that if a homothecy preserves distance, then it is either the identity or harmonic.

17. A *shear* is an elation on π_R whose axis is finite and whose center is ideal. Investigate properties of shears.

18. A *glide* is a homology on π_R whose axis is finite and whose center is ideal. Investigate properties of glides.

19. Show that reflection in a line is a harmonic glide.

20. Show that the composition of two reflections in parallel lines is a translation.

21. Show that the composition of two reflections in intersecting lines is a rotation.

Section 2.7. Automorphic Collineations on π_F

In this section we consider another type of collineation on the plane over a field, this time induced by a certain function on the field.

Let F be a field. An *automorphism* of F is a bijection $\varphi : F \to F$ satisfying

1. $\varphi(x+y) = \varphi(x) + \varphi(y)$, and
2. $\varphi(xy) = \varphi(x)\varphi(y)$.

That is, an automorphism of F is a ring isomorphism from F onto itself. Two properties of an automorphism φ on F are $\varphi(0) = 0$ and $\varphi(1) = 1$.

If φ is an automorphism of field F, we define a function $f_\varphi : \pi_F \to \pi_F$ by

$$f_\varphi[x_1, x_2, x_3] = [\varphi(x_1), \varphi(x_2), \varphi(x_3)].$$

Theorem 1. *f_φ is a collineation.*

PROOF. First we show that f_φ is one-to-one. If $f_\varphi[x_1, x_2, x_3] = f_\varphi[y_1, y_2, y_3]$, then $[\varphi(x_1), \varphi(x_2), \varphi(x_3)] = [\varphi(y_1), \varphi(y_2), \varphi(y_3)]$. Hence for some $k \in F$, $\varphi(x_i) = k\varphi(y_i)$, $i = 1, 2, 3$. Hence $x_i = \varphi^{-1}(k)y_i$, $i = 1, 2, 3$, and $[x_1, x_2, x_3] = [y_1, y_2, y_3]$. Thus f_φ is one-to-one. Next, to show that f_φ is onto, let $[y_1, y_2, y_3]$ be any point of π_F. Then also $[\varphi^{-1}(y_1), \varphi^{-1}(y_2), \varphi^{-1}(y_3)]$ is a point of π_F, and $f_\varphi[\varphi^{-1}(y_1), \varphi^{-1}(y_2), \varphi^{-1}(y_3)] = [y_1, y_2, y_3]$. Thus f_φ is onto. Finally, to show that f_φ preserves collinearity, let $x_1 = [x_{11}, x_{12}, x_{13}]$, $x_2 = [x_{21}, x_{22}, x_{23}]$, and $x_3 = [x_{31}, x_{32}, x_{33}]$ be three points. Then $f_\varphi(x_1)$, $f_\varphi(x_2)$, and $f_\varphi(x_3)$ are collinear if and only if $\det(\varphi(x_{ij})) = 0$, if and only if $\varphi(\det(x_{ij})) = 0$, if and only if $\det(x_{ij}) = 0$, if and only if x_1, x_2, and x_3 are collinear. □

We call f_φ an *automorphic collineation* of π_F. The following theorem highlights the contrast between projective and automorphic collineations.

Theorem 2. *No automorphic collineation but the identity is projective.*

PROOF. Suppose f_φ is an automorphic collineation that is projective. On line $\langle 1,0,0 \rangle$, points $[0,1,0]$, $[0,0,1]$, and $[0,1,1]$ are fixed by f_φ. Since $f_\varphi|_{\langle 1,0,0\rangle}$ is a projectivity, it follows from Exercise 2.2.6 that $f_\varphi|_{\langle 1,0,0\rangle}$ is the identity, or $\langle 1,0,0 \rangle$ is pointwise fixed by f_φ. Similarly, on line $\langle 0,1,0 \rangle$, the points $[1,0,0]$, $[0,0,1]$, and $[1,0,1]$ are fixed by f_φ, so line $\langle 0,1,0 \rangle$ is pointwise fixed. Hence f_φ is the identity by Theorem 2.3.1. □

A collineation $f : \pi_F \to \pi_F$ will be called *basic* in case f fixes the points

$[1,0,0]$, $[0,1,0]$, $[0,0,1]$, and $[1,1,1]$. It should be obvious that every automorphic collineation is basic. But the converse is not so easily seen.

Theorem 3. *Every basic collineation on π_F is automorphic.*

PROOF. Let $f: \pi_F \to \pi_F$ be a basic collineation. Then f fixes points $[1,0,0]$ and $[0,1,0]$, so line $\langle 0,0,1 \rangle$ is fixed by f. Every point of the form $[1,x,0]$ thus has image on line $\langle 0,0,1 \rangle$; moreover, its image is of the form $[1,y,0]$, for $[1,x,0] \neq [0,1,0]$. Hence we define $\varphi: F \to F$ by

$$f[1,x,0] = [1,\varphi(x),0].$$

By the foregoing, φ is well defined. To show that φ is one-to-one, suppose $\varphi(x_1) = \varphi(x_2)$. Then $f[1,x_1,0] = [1,\varphi(x_1),0] = [1,\varphi(x_2),0] = f[1,x_2,0]$. But since f is one-to-one, $[1,x_1,0] = [1,x_2,0]$, so $x_1 = x_2$. Thus φ is one-to-one. To show that φ is onto, let $y \in F$. Then, since f is onto, there is a point $[1,x,0]$ such that $f[1,x,0] = [1,y,0]$. Hence $\varphi(x) = y$, and φ is onto. We also have $\varphi(0) = 0$, for $f[1,0,0] = [1,0,0]$. Also, point $[1,1,0]$ is collinear with both $[1,0,0]$ and $[0,1,0]$ and also collinear with both $[0,0,1]$ and $[1,1,1]$. Hence $f[1,1,0] = [1,1,0]$, and $\varphi(1) = 1$. By computing coordinates (see Figure 2.10), we can see that $\varphi(x_1 + x_2) = \varphi(x_1) + \varphi(x_2)$. In a similar way (see Figure 2.11), we can see that $\varphi(x_1 x_2) = \varphi(x_1)\varphi(x_2)$. Hence φ is an automorphism.

It remains to be shown that f is the same as f_φ. Clearly f and f_φ are the same on line $\langle 0,0,1 \rangle$, for $f[1,x,0] = [1,\varphi(x),0] = [\varphi(1),\varphi(x),\varphi(0)] = f_\varphi[1,x,0]$ and $f[0,1,0] = f_\varphi[0,1,0] = [0,1,0]$. We next show that f and f_φ are the same on line $\langle 0,1,-1 \rangle$ (Figure 2.12). If p is any point on $\langle 0,1,-1 \rangle$, then p has coordinates $[x_1,x_2,x_2]$. Let q be the point $[x_1,x_2,0]$; then p, q, and $[0,0,1]$ are collinear. If $x_1 = 0$, then

Figure 2.10

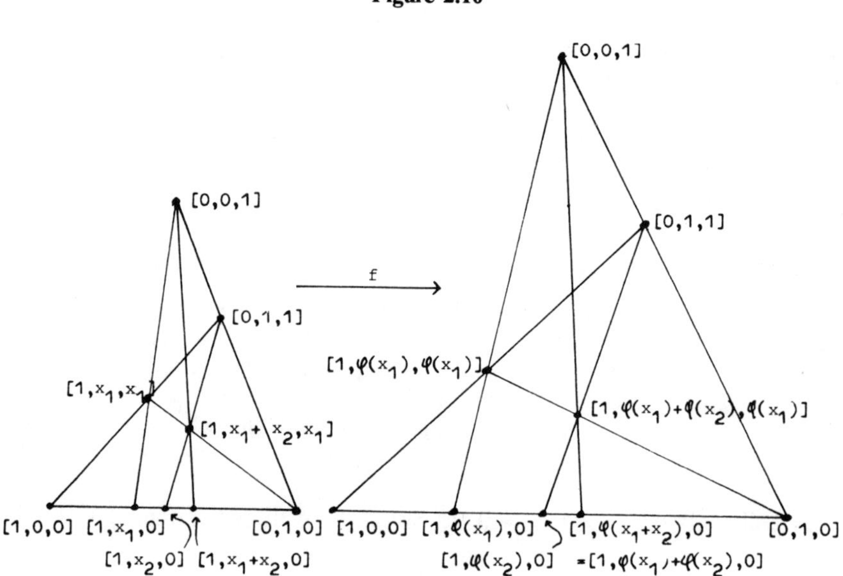

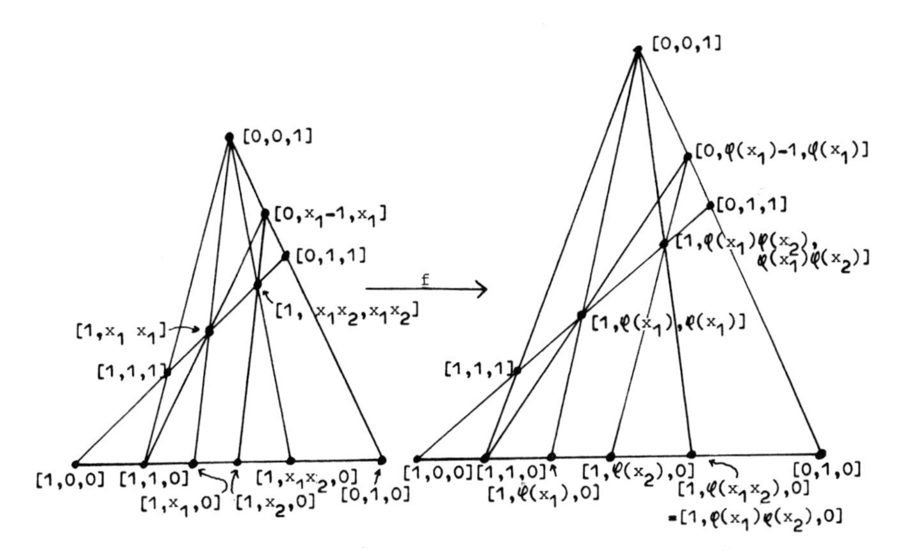

Figure 2.11

$p = [0, 1, 1]$ and $f(p) = f_\varphi(p)$. If $x_1 \neq 0$, then $q = [1, x_1^{-1}x_2, 0]$ and $f(q) = [1, \varphi(x_1^{-1}x_2), 0] = [\varphi(x_1), \varphi(x_2), 0]$. Hence $f(p)$ is on both $\langle 0, 1, -1\rangle$ and the line on $[0, 0, 1]$ and $[\varphi(x_1), \varphi(x_2), 0]$. Thus $f(p) = [\varphi(x_1), \varphi(x_2), \varphi(x_3)] = f_\varphi(p)$. Thus f and f_φ are the same on $\langle 0, 1, -1\rangle$. Finally, let p be a point not on $\langle 0, 0, 1\rangle$ or $\langle 0, 1, -1\rangle$. Then $p = [x_1, x_2, x_3]$ with $x_2 \neq x_3$ and $x_3 \neq 0$. The line on p and $[1, 1, 1]$ (see Figure 2.13) is $\langle x_3 - x_2, x_1 - x_3, x_2 - x_1\rangle$, which meets $\langle 0, 0, 1\rangle$ at $q = [x_1 - x_3, x_2 - x_3,$

Figure 2.12

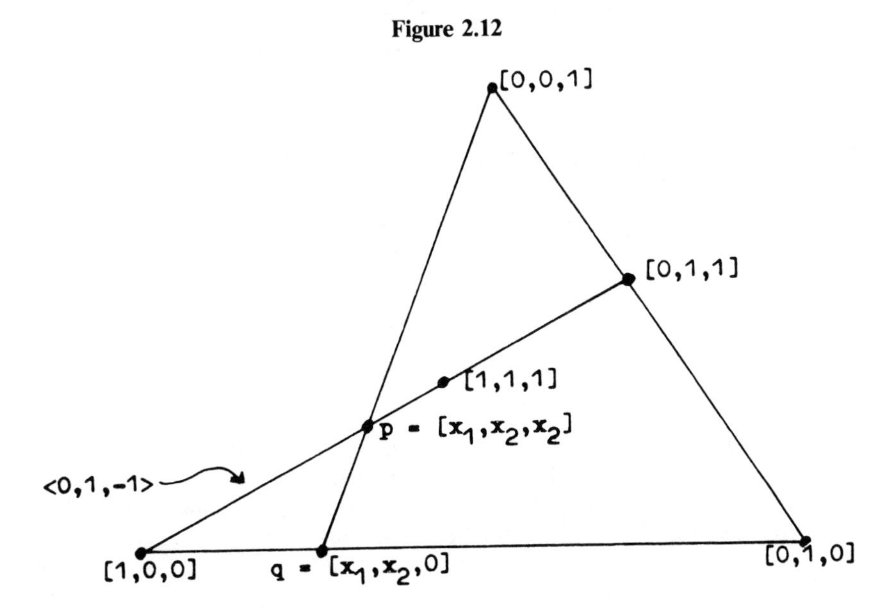

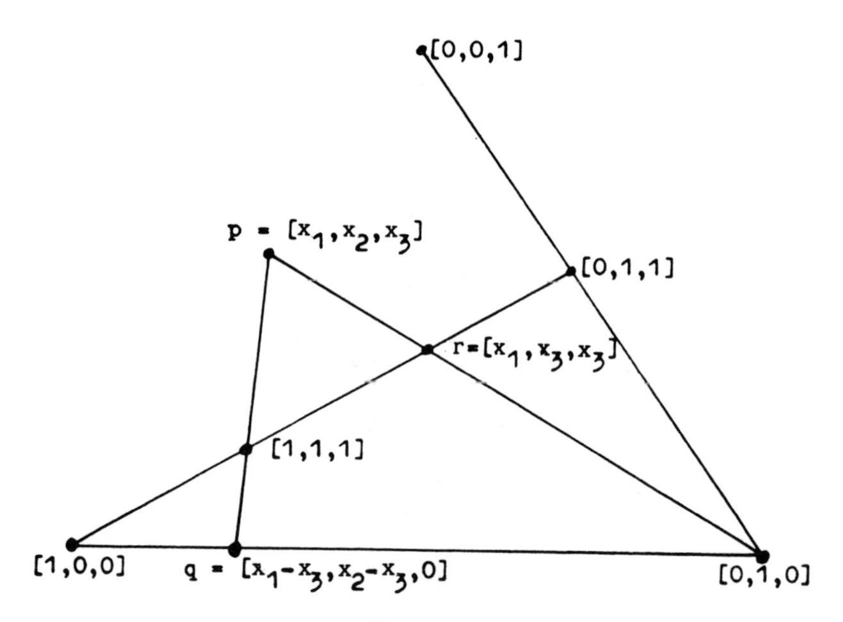

Figure 2.13

0]. The line on p and $[0, 1, 0]$ is $\langle x_3, 0, -x_1\rangle$, which meets $\langle 0, 1, -1\rangle$ at $r = [x_1, x_3, x_3]$. Hence $f(q) = f_\varphi(q)$ and $f(r) = f_\varphi(r)$, so $f(p)$ is on both the line joining $f_\varphi(q)$ to $[1, 1, 1]$ and the line joining $f_\varphi(r)$ to $[0, 1, 0]$. Thus $f(p)$ is $f_\varphi(p)$, and we have shown $f = f_\varphi$. Thus every basic collineation is automorphic. □

We next wish to consider automorphic collineations on π_R. However, there are none but the identity, as the next theorems show.

Theorem 4. *The only automorphism of the real number field R is the identity.*

PROOF. Let $\varphi : R \to R$ be an automorphism. Then $\varphi(0) = 0$ and $\varphi(1) = 1$. If n is a positive integer, then

$$\varphi(n) = \varphi(\underbrace{1 + 1 + \cdots + 1}_{n \text{ times}})$$

$$= \underbrace{\varphi(1) + \varphi(1) + \cdots + \varphi(1)}_{n \text{ times}}$$

$$= \underbrace{1 + 1 + \cdots + 1}_{n \text{ times}} = n,$$

and $\varphi(-n) = -\varphi(n) = -n$. Hence if n is any integer, $\varphi(n) = n$. Also, $\varphi(n^{-1}) = \varphi(n)^{-1}$, so if m/n is any rational number, then $\varphi(m/n) = \varphi(mn^{-1}) = \varphi(m)\varphi(n)^{-1} = m/n$. Thus if r is any rational number, then $\varphi(r) = r$. Now suppose x is a positive real number. Then there is another positive real number y such that $x = y^2$. Hence $\varphi(x) = \varphi(y^2) = \varphi(y)^2$. Since $\varphi(y) \neq 0$, $\varphi(x)$ must also be positive. That is, if $x > 0$, then $\varphi(x) > 0$. Hence if $x < y$, then $y - x > 0$, so $\varphi(y) - \varphi(x) = \varphi(y - x)$

> 0 and $\varphi(x) < \varphi(y)$. Thus φ preserves order. Let x be any real number. If $\varphi(x) < x$, then there is a rational number r such that $\varphi(x) < r < x$. But then $\varphi(r) < \varphi(x)$, or $r < \varphi(x)$, a contradiction. Similarly, if $\varphi(x) > x$, then there is a rational number s such that $x < s < \varphi(x)$. But then $\varphi(x) < \varphi(s) = s$, a contradiction. Hence it must be that $\varphi(x) = x$, and φ is the identity. □

Theorem 5. *Every collineation on π_R is projective.*

PROOF. Let $f: \pi_R \to \pi_R$ be a collineation. Let $f[1,0,0] = p$, $f[0,1,0] = q$, $f[0,0,1] = r$, and $f[1,1,1] = s$. Now let f_M be a matrix-induced collineation carrying $[1,0,0]$ to p, $[0,1,0]$ to q, $[0,0,1]$ to r, and $[1,1,1]$ to s (see Exercise 2.4.8). Then $f^{-1} \cdot f_M$ is a basic collineation, so by Theorem 3 it is automorphic, $f^{-1} \cdot f_M = f_\varphi$. But on R, φ must be the identity, so f_φ is the identity and $f^{-1} \cdot f_M$ is the identity. Hence $f = f_M$, so it is projective (Theorem 2.4.3). □

Techniques similar to this last proof allow us to identify any projective collineation on the plane over a field as a matrix-induced collineation:

Theorem 6. *Every projective collineation on π_F is matrix induced.*

PROOF. Let $f: \pi_F \to \pi_F$ be a projective collineation. Let $f[1,0,0] = p$, $f[0,1,0] = q$, $f[0,0,1] = r$, and $f[1,1,1] = s$. Let f_M be a matrix-induced collineation carrying $[1,0,0]$ to p, $[0,1,0]$ to q, $[0,0,1]$ to r, and $[1,1,1]$ to s. Then $f^{-1} \cdot f_M$ is basic, and hence automorphic. Thus $f^{-1} \cdot f_M = f_\varphi$ for some automorphism φ. But since f and f_M are both projective, it follows that f_φ is projective. Hence by Theorem 2, f_φ is the identity. Thus $f = f_M$, and f is matrix induced. □

Exercises 2.7

1. Is the set of automorphic collineations on π_F a group?

2. Prove that an automorphic collineation is basic.

3. Let $f, g: \pi_F \to \pi_F$ be collineations. We say that f is *equivalent* to g, $f \simeq g$, if and only if there exists projective collineations h_1 and h_2 on π_F such that $f = h_2 \cdot g \cdot h_1$. Prove that $\simeq$ is an equivalence relation.

4. Prove that any matrix-induced collineation on π_F is equivalent to the identity.

5. Prove that there is a one-to-one correspondence between the automorphisms of F and the $\simeq$-equivalences classes of collineations on π_F.

Chapter 3

Desarguesian and Pappian Planes

In this chapter we develop the properties of some particular classes of projective planes. These planes are characterized by certain properties which were found to hold (in a projective sense) in the Euclidean plane. They are named after the men who discovered those properties. Thus, the planes we study here are in a sense generalizations of the real projective plane.

Section 3.1. Desarguesian Planes

In 1639, the Frenchman Gerard Desargues (d'zarg′) published a book that investigated the real projective plane (though, of course, the real projective plane was not known by that name until almost three centuries later). One of the theorems he proved is known as Desargues's triangle theorem, which we will state below. Any projective plane in which Desargues's triangle theorem holds is called a *Desarguesian plane*.

Two triangles, $\triangle abc$ and $\triangle a'b'c'$ are said to form a *couple* in case the points a, b, c, a', b', c' are six distinct points and the lines aa', bb', cc' are three distinct lines, none of which is a side of either triangle. The couple $\triangle abc$ and $\triangle a'b'c'$ is said to be *central*, with *center* p, in case the lines aa', bb', and cc' are concurrent on p (see Figure 3.1). The couple is said to be *axial*, with *axis* L, in case the points $ab \cap a'b'$, $ac \cap a'c'$, and $bc \cap b'c'$ are collinear on L (see Figure 3.2).

We can now state Desargues's triangle theorem as follows:

Desargues's triangle theorem. *Every central couple is axial.*

Thus a Desarguesian plane is a projective plane in which every central couple is axial (see Figure 3.3).

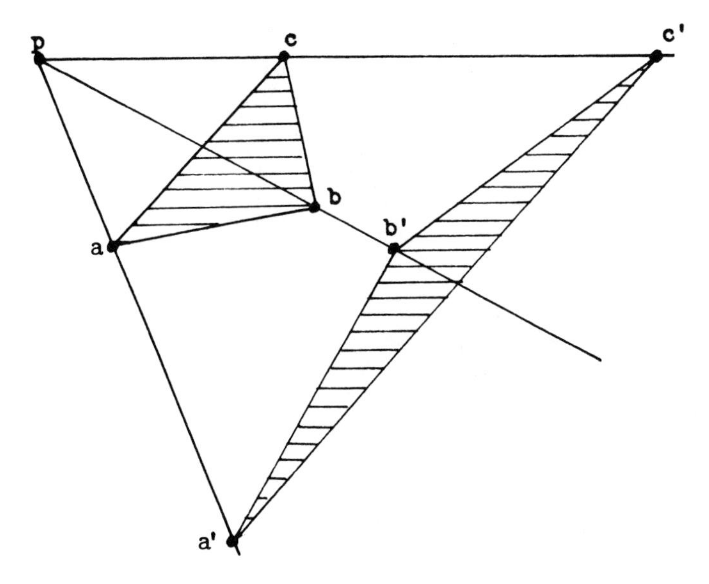

Figure 3.1

We first verify that the real projective plane is Desarguesian, by proving a more general result.

Theorem 1. *If D is a division ring, then π_D is a Desarguesian plane.*

PROOF. As a notational convention, we let point $x \in \pi_D$ have coordinates denoted by $[x_1, x_2, x_3]$. Now let $\triangle abc$ and $\triangle a'b'c'$ form a central couple, with aa', bb', and

Figure 3.2

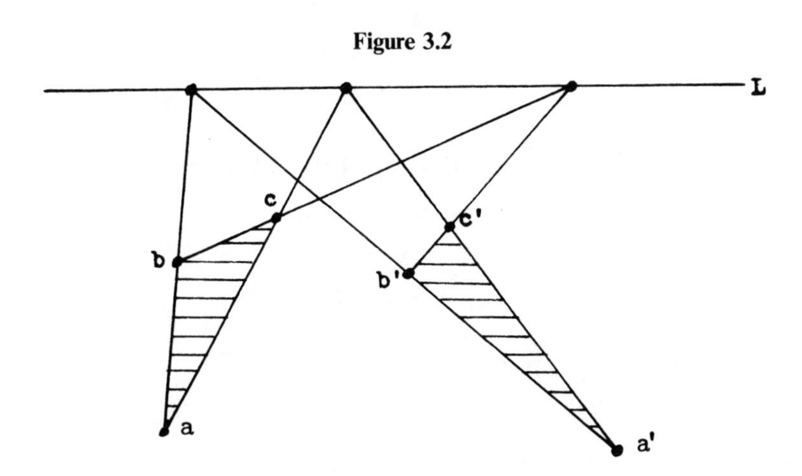

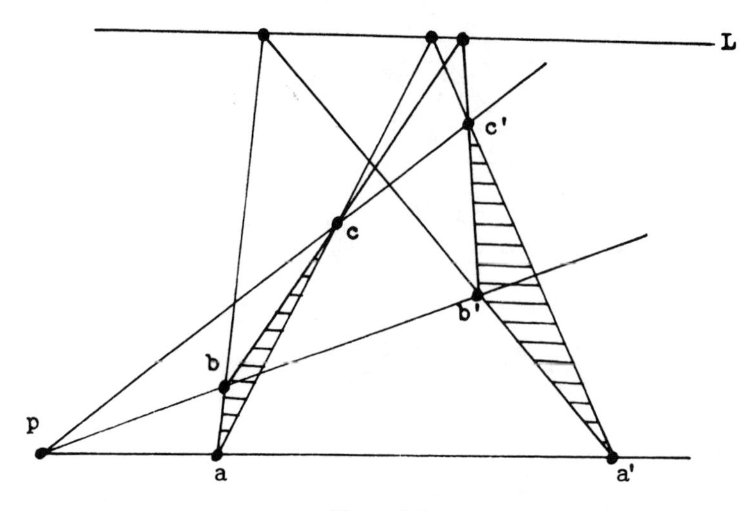

Figure 3.3

cc' concurrent on p. Then there exist $\alpha,\ \beta, \gamma \in D$ such that for each $i = 1, 2, 3$,

$$a_i' = p_i + \alpha a_i,$$
$$b_i' = p_i + \beta b_i,$$
$$c_i' = p_i + \gamma c_i.$$

Let $a'' = bc \cap b'c'$, $b'' = ac \cap a'c'$, and $c'' = ab \cap a'b'$. Since a'' is on bc, there exists $\lambda \in D$ such that

$$a_i'' = b_i + \lambda c_i.$$

Since a'' is on $b'c'$, there exist $k, \lambda' \in D$ such that

$$\begin{aligned} a_i'' &= k(b_i' + \lambda' c_i') \\ &= k[p_i + \beta b_i + \lambda'(p_i + \gamma c_i)] \\ &= k(1 + \lambda')p_i + k\beta b_i + k\lambda'\gamma c_i. \end{aligned}$$

Hence

$$b_i + \lambda c_i = k(1 + \lambda')p_i + k\beta b_i + k\lambda'\gamma c_i,$$

or

$$k(1 + \lambda')p_i + (k\beta - 1)b_i + (k\lambda'\gamma - \lambda)c_i = 0.$$

But since p, b, and c are not collinear, we must have

$$k(1 + \lambda') = 0,$$
$$k\beta - 1 = 0,$$
$$k\lambda'\gamma - \lambda = 0.$$

Solving these equations, we find

$$k = \beta^{-1}, \qquad \lambda' = -1, \qquad \lambda = -\beta^{-1}\gamma.$$

Hence

$$a_i'' = b_i - \beta^{-1}\gamma c_i.$$

In a similar fashion, we find

$$b_i'' = c_i - \gamma^{-1}\alpha a_i$$

and

$$c_i'' = a_i - \alpha^{-1}\beta b_i.$$

Hence

$$\beta a_i'' + \gamma b_i'' + \alpha c_i'' = 0,$$

so a'', b'', and c'' are collinear by Exercise 1.8.15. Hence $\triangle abc$ and $\triangle a'b'c'$ form an axial couple. □

It follows from Theorem 1 that if F is a field, then π_F is a Desarguesian plane. In particular, π_R is a Desarguesian plane.

To prove that the principle of duality holds in the class of Desarguesian planes, it is sufficient to prove that the dual of Desargues's triangle theorem holds in a Desarguesian plane. The dual turns out to be the converse (see the exercises). Hence, we prove the converse.

Theorem 2. *If π is a Desarguesian plane, then every axial couple in π is central.*

PROOF. Let $\triangle abc$ and $\triangle a'b'c'$ form an axial couple, with $a'' = bc \cap b'c'$, $b'' = ac \cap a'c'$, and $c'' = ab \cap a'b'$ collinear on line L (see Figure 3.4). Let $p = bb' \cap cc'$. Now $bb'c''$ and $cc'b''$ form a central couple with center a''. Hence the points

$$bb' \cap cc' = p,$$
$$bc'' \cap cb'' = a,$$
$$b'c'' \cap c'b'' = a'$$

are collinear. Hence aa' is on p, and $\triangle abc$ and $\triangle a'b'c'$ form a central couple. □

Recall that a *four-point* is a set of four points, no three of which are collinear, and a *complete four-point* is a four-point together with the six lines, called *sides*, determined by pairs of the four points. Two sides that do not meet in one of the four points are called *opposite* sides, and meet in a point called a *diagonal point*. Thus, associated with a four-point in a projective plane are three diagonal points.

A projective plane in which every four-point has collinear diagonal points is called a *Fano plane*. As the extreme opposite, a projective plane in which *no* four-point has collinear diagonal points is said, by a quirk of history, to satisfy *Fano's axiom*.

A Desarguesian plane either is a Fano plane or satisfies Fano's axiom, as the following theorems show.

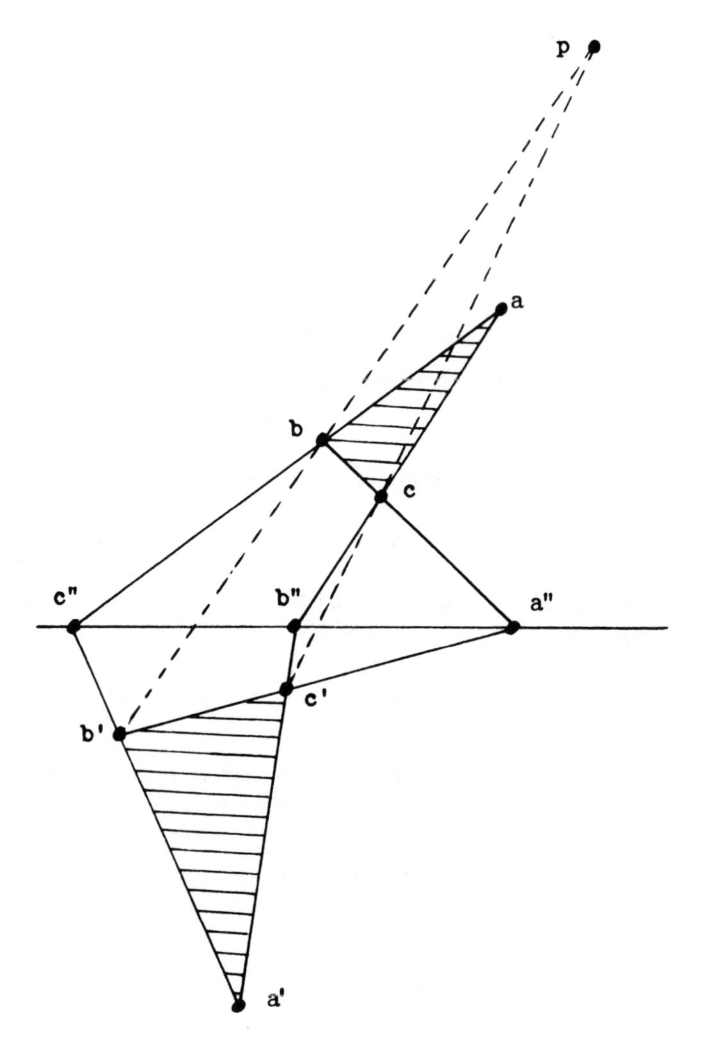

Figure 3.4

Theorem 3. *Let $a_0a_1a_2a_3$ and $b_0b_1b_2b_3$ be two four-points in a Desarguesian plane π such that $a_2 = b_2$, $a_3 = b_3$, and $a_0a_1 \cap a_2a_3 = b_0b_1 \cap b_2b_3$. Then the diagonal points of $a_0a_1a_2a_3$ are collinear if and only if the diagonal points of $b_0b_1b_2b_3$ are collinear.*

PROOF. Let $d_1 = a_0a_1 \cap a_2a_3 = b_0b_1 \cap b_2b_3 = e_1$, $d_2 = a_0a_2 \cap a_1a_3$, $d_3 = a_0a_3 \cap a_1a_2$, $e_2 = b_0b_2 \cap b_1b_3$, and $e_3 = b_0b_3 \cap b_1b_2$ (see Figure 3.5). Then $\triangle a_0a_1d_2$ and $\triangle b_0b_1e_2$ form an axial couple, and hence a central couple, so that lines a_0b_0, a_1b_1, and d_2e_2 are concurrent. In addition, $\triangle a_0a_1d_3$ and $\triangle b_0b_1e_3$ form an axial couple, so that lines a_0b_0, a_1b_1, and d_3e_3 are concurrent. Hence lines a_0b_0, d_2e_2, and d_3e_3 are concurrent, so $\triangle a_0d_2d_3$ and $\triangle b_0e_2e_3$ form a central couple, and hence an axial

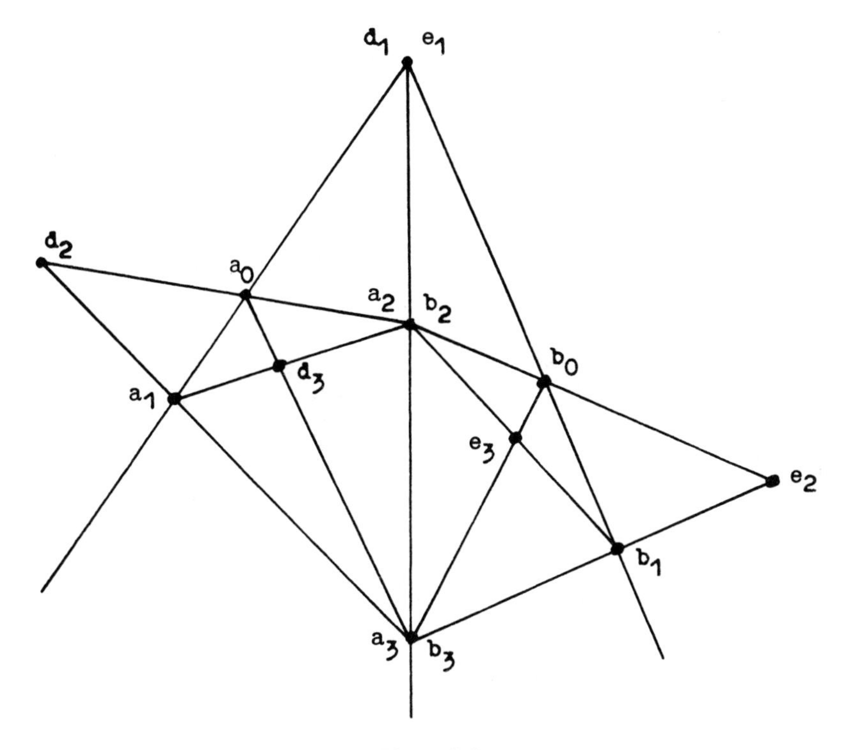

Figure 3.5

couple. Thus $a_0d_2 \cap b_0e_2 = a_2 = b_2$, $a_0d_3 \cap b_0e_3 = a_3 = b_3$, and $d_2d_3 \cap e_2e_3$ are collinear points. Now suppose d_1, d_2, d_3 are collinear. Then $d_2d_3 \cap a_2a_3 = d_1$; but $d_2d_3 \cap a_2a_3 = d_2d_3 \cap e_2e_3 = e_2e_3 \cap b_2b_3$, so e_2e_3 meets b_2b_3 in $d_1 = e_1$, and e_1, e_2, and e_3 are also collinear. The converse is similar. □

Theorem 4. *In a Desarguesian plane π, if one four-point has collinear diagonal points, then every four-point has collinear diagonal points.*

PROOF. Let $a_0a_1a_2a_3$ be a four-point in π which has collinear diagonal points, and let $b_0b_1b_2b_3$ be an arbitrary four-point in π. Let $d = a_0a_1 \cap a_2a_3$, $e = b_0b_1 \cap b_2b_3$, $f = a_0b_1 \cap a_1b_0$, $g = de \cap a_0b_1$, and $h = de \cap a_1b_0$ (see Figure 3.6). Then we may apply Theorem 3 to each of the following pairs of four-points:

$$a_2a_3a_0a_1 \quad \text{and} \quad gha_0a_1,$$
$$a_1ha_0g \quad \text{and} \quad hb_0a_0g,$$
$$ghb_0b_1 \quad \text{and} \quad b_2b_3b_0b_1.$$

Hence $b_0b_1b_2b_3$ also has collinear diagonal points. □

Some modifications of the above proof are necessary for the several possible special cases; they are left to the reader.

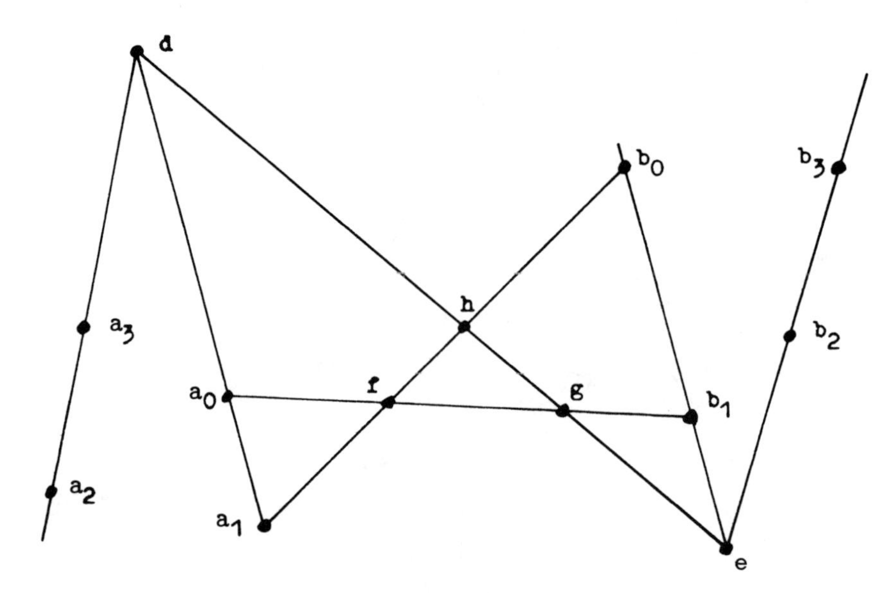

Figure 3.6

Let *abcd* be a complete four-point, and let L be a line not on a, b, c, or d. Then L meets the sides of *abcd* in six (not necessarily distinct) points. The set of three pairs of points in which L meets each pair of opposite sides of *abcd* is called the *quadrangular set* on L *induced* by *abcd*.

For example, if $a_0a_1a_2a_3$ is a four-point, L is a line not on any a_i, and

$$q_1 = a_0a_1 \cap L, \qquad q_1' = a_2a_3 \cap L,$$
$$q_2 = a_0a_2 \cap L, \qquad q_2' = a_1a_3 \cap L,$$
$$q_3 = a_0a_3 \cap L, \qquad q_3' = a_1a_2 \cap L,$$

then

$$\{(q_1, q_1'), (q_2, q_2'), (q_3, q_3')\}$$

is a quadrangular set (see Figure 3.7).

Theorem 5. *If p, q, r, s, t are five collinear points in a projective plane π, then there is a point u such that*

$$\{(p, q), (r, s), (t, u)\}$$

is a quadrangular set.

PROOF. Let p, q, r, s, t lie on line L, and let a be a point not on L (see Figure 3.8). Let b be a third point on ap. Let

$$d = at \cap bs,$$
$$c = ar \cap dq,$$
$$u = bc \cap L.$$

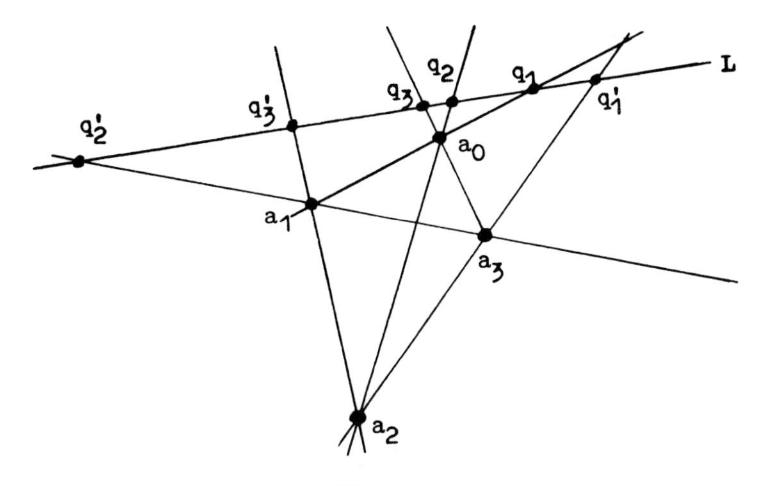

Figure 3.7

Then $p = ab \cap L$, $q = cd \cap L$; $r = ac \cap L$, $s = bd \cap L$; and $t = ad \cap L$, $u = bc \cap L$. Hence $\{(p, q), (r, s), (t, u)\}$ is a quadrangular set. □

Theorem 6. *Point u in Theorem 5 is unique if and only if π is Desarguesian.*

PROOF. We first assume that π is Desarguesian. To show that u is unique, suppose that u has been constructed as in the proof of Theorem 5, once using four-point $abcd$, and a second time using four-point $a'b'c'd'$ (see Figure 3.9). We wish to show that $b'c'$ also meets L at u, just as bc does, so that the second construction yields the same point u. To do so, we notice that $\triangle abd$ and $\triangle a'b'd'$ form an axial couple, so that aa', bb', and dd' are concurrent. Also, $\triangle acd$ and $\triangle a'c'd'$ form an axial couple,

Figure 3.8

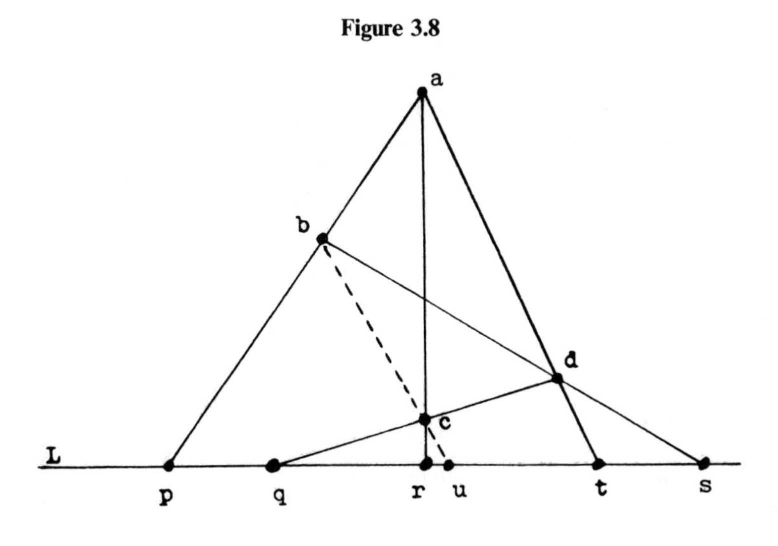

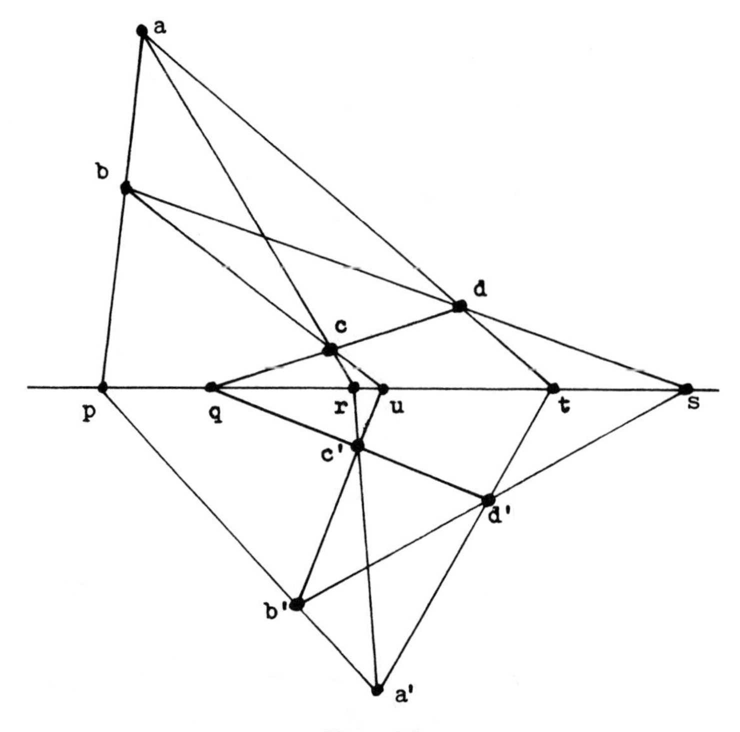

Figure 3.9

so that aa', cc', and dd' are concurrent. Hence aa', bb', and cc' are concurrent, so that $\triangle abc$ and $\triangle a'b'c'$ form a central couple. Hence $ab \cap a'b' = p$, $ac \cap a'c' = r$, and $bc \cap b'c'$ are collinear. That is, $bc \cap b'c'$ is on L; but $bc \cap L = u$, so $b'c' \cap L = u$ also, and u is unique.

Conversely, let $\triangle abc$ and $\triangle a'b'c'$ form a central couple with center o (see Figure 3.10). Let $c'' = ab \cap a'b'$, $b'' = ac \cap a'c'$, and $L = b''c''$. Let $p = aa' \cap L$, $q = bb' \cap L$, and $r = cc' \cap L$. Let $x = bc \cap L$ and $x' = b'c' \cap L$. Finally, let $x = bc \cap L$ and $x' = b'c' \cap L$. We must show $x = x'$. Now $\{(q, b'')(r, c''), (p, x)\}$ is the quadrangular set on L induced by the four-point $oabc$, and $\{(q, b''), (r, c''), (p, x')\}$ is the quadrangular set induced on L by the four-point $oa'b'c'$. Since we are assuming that the sixth quadrangular point is unique, we must have $x = x'$. Hence π is Desarguesian. □

Exercises 3.1

1. Show that the dual of a couple is a couple, and the dual of a central couple is an axial couple.

*2. Prove that the principle of duality holds in the class of Desarguesian planes.

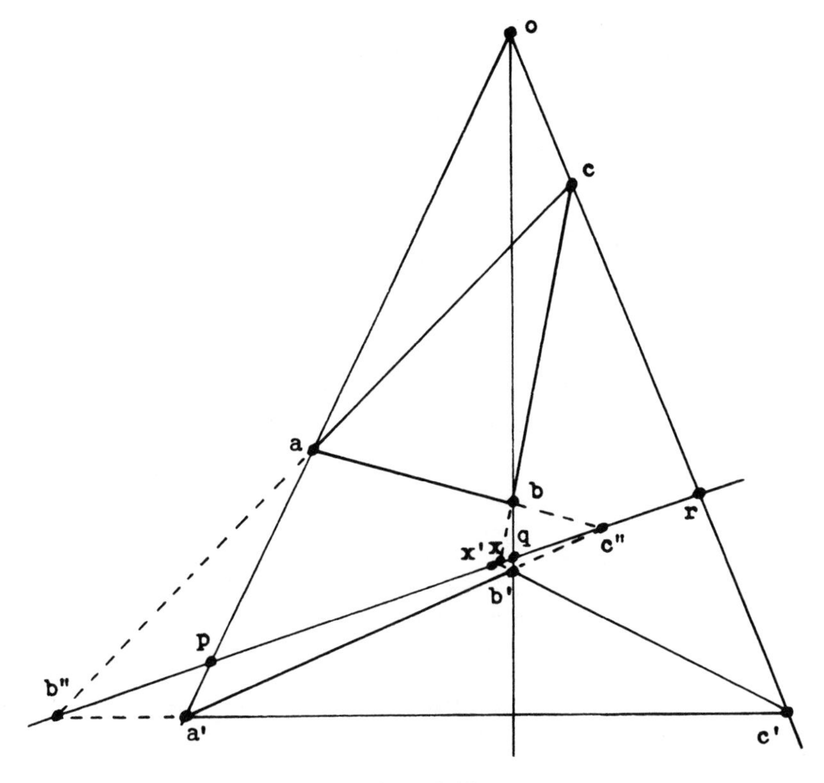

Figure 3.10

3. Use Desargues's triangle theorem to prove tnat the medians of a triangle in α_R are concurrent.

4. Show that the Veblen–Wedderburn plane π_V (Example 1.8.4) is not a Fano plane yet does not satisfy Fano's axiom.

*5. The points of a quadrangular set are not always six distinct points. Discuss all possible cases.

*6. In Theorem 5, the given points p, q, r, s, t need not be five distinct points. What distinctions are necessary in order for the theorem to hold?

*7. If line L lies on two diagonal points g and h of four-point $abcd$, the quadrangular set $\{(g, g), (h, h)(x, y)\}$ on L induced by $abcd$ is called a *harmonic set*, and is denoted by (gh, xy). The points x and y are called *harmonic conjugates* of each other relative to g and h. Explain why, in a Desarguesian plane, the harmonic conjugate of x relative to g and h is unique.

***8.** Given three collinear points g, h, and x in a Desarguesian plane, show how to construct the harmonic conjugate of x relative to g and h.

***9.** Let π be a Desarguesian plane. Prove that π is a Fano plane if and only if for every harmonic set (ab, cd) we have $c = d$.

10. Let π be a Desarguesian plane satisfying Fano's axiom. Let $a_0a_1a_2a_3$ be a four-point in π, with diagonal points

$$d_1 = a_0a_1 \cap a_2a_3, \quad d_2 = a_3a_2 \cap a_1a_3, \quad d_3 = a_0a_3 \cap a_1a_2.$$

The six points

$$h_1 = a_0a_1 \cap d_2d_3, \quad h_1' = a_2a_3 \cap d_2d_3,$$
$$h_2 = a_0a_2 \cap d_1d_3, \quad h_2' = a_1a_3 \cap d_1d_3,$$
$$h_3 = a_0a_3 \cap d_1d_2, \quad h_3' = a_1a_2 \cap d_1d_2$$

are called *harmonic points* of $a_0a_1a_2a_3$. Verify that (d_1d_2, h_3h_3'), (d_1d_3, h_2h_2'), and (d_2d_3, h_1h_1') are harmonic sets.

***11.** In the notation of Exercise 10, show that h_1', h_2, h_3 are collinear, h_1, h_2', h_3 are collinear, h_1, h_2, h_3' are collinear, and h_1', h_2', h_3' are collinear. That is, the harmonic points of a four-point are the vertices of a four-line. (A *four-line* is the dual of a four-point, and a vertex is the dual of a side.)

***12.** Prove that in a Desarguesian plane satisfying Fano's axiom, if (ab, cd) is a harmonic set, then (cd, ab) is also a harmonic set.

13. If (ab, cd) is a harmonic set in a Desarguesian plane satisfying Fano's axiom, show that the following are also harmonic sets: (ab, dc), (ba, cd), (ba, dc), (cd, ab), (cd, ba), (dc, ab), (dc, ba).

14. Show that a Fano configuration cannot be embedded in π_R or in α_R.

Section 3.2. Projectivities in Desarguesian Planes

Our first objective in this section is to prove that a projectivity in a Desarguesian plane can always be expressed as a composition of at most three perspectivities. Some rather technical theorems lead to that result.

All the theorems in this section are for Desarguesian planes.

Theorem 1. *If L_1, L_2, and L_3 are three concurrent lines, and $f: L_1 \overset{u}{\barwedge} L_2$ and $g: L_2 \overset{v}{\barwedge} L_3$ are perspectivities, then $g \cdot f$ is a perspectivity, and for some point w on uv, $g \circ f: L_1 \underset{w}{\barwedge} L_3$.*

PROOF. Let p_1 be a point of L_1, not on L_2 or on uv. Let $p_2 = f(p_1)$, $p_3 = g(p_2)$. Let x_1 be an arbitrary point on L_1; let $x_2 = f(x_1)$ and $x_3 = g(x_2)$. Let d be the point of concurrency of L_1, L_2, and L_3 (see Figure 3.11). Now $\triangle p_1p_2p_3$ and $\triangle x_1x_2x_3$ form a central couple with center d. Hence points $p_1p_2 \cap x_1x_2 = u$, $p_2p_3 \cap x_1x_3 = v$, and

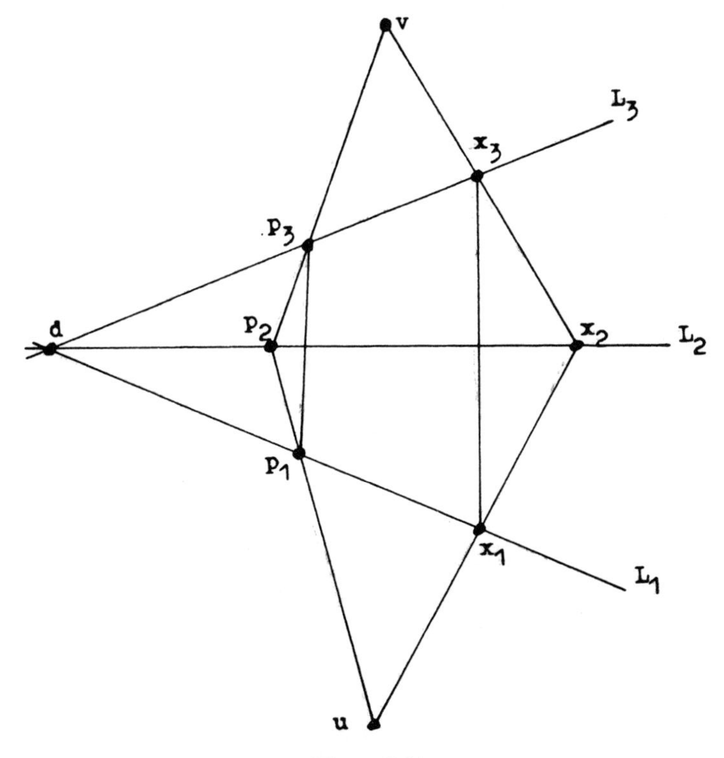

Figure 3.11

$p_1p_3 \cap x_1x_3$ are collinear. Let $w = p_1p_3 \cap x_1x_3$. Then w is on uv, and since $g \circ f(x_1) = x_3$, $g \circ f$ is a perspectivity with center w. □

Theorem 2. *Let L_1, L_2, L_3 be three nonconcurrent lines, and let $f : L_1 \overset{u}{\barwedge} L_2$ and $g : L_2 \overset{v}{\barwedge} L_3$ be perspectivities. If L_2' is a third line on $L_1 \cap L_2$, but not on v, then there exist perspectivities $f' : L_1 \overset{w}{\barwedge} L_2'$, with w on uv, and $g' : L_2' \overset{v}{\barwedge} L_3$ such that $g' \circ f' = g \circ f$.*

PROOF. Let g_1 be the perspectivity

$$g_1 \cdot L_2 \overset{v}{\barwedge} L_2'.$$

Then by Theorem 1, $g_1 \circ f$ is a perspectivity with center w on uv (see Figure 3.12). Let $f' = g_1 \circ f$; then

$$f' : L_1 \overset{w}{\barwedge} L_2'.$$

Let g' be the perspectivity

$$g' : L_2' \overset{v}{\barwedge} L_3.$$

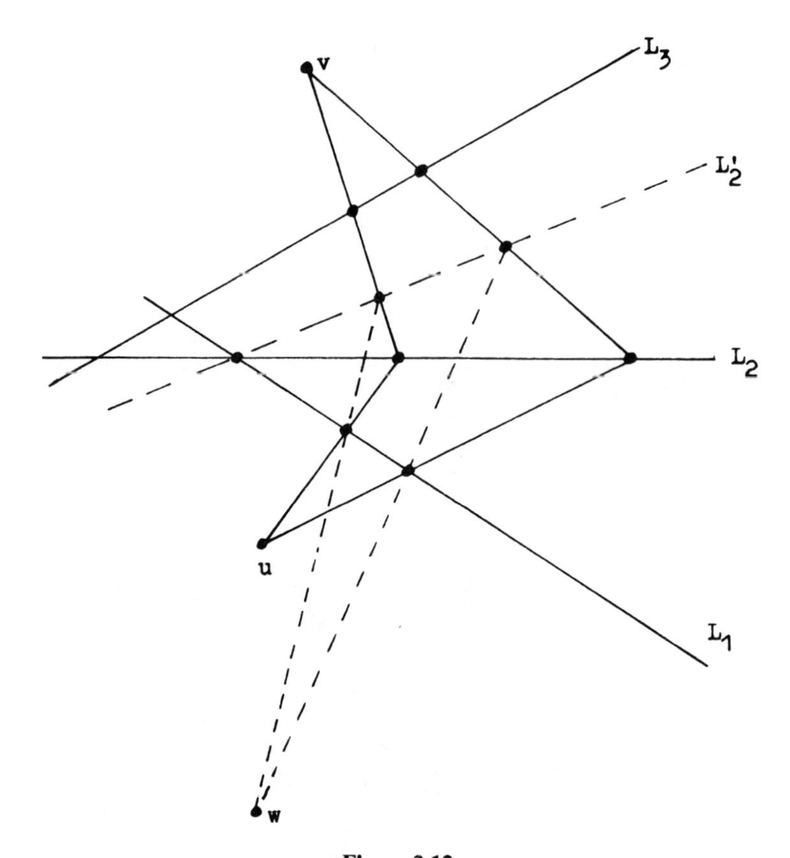

Figure 3.12

Then $g' \circ g_1 = g$, so

$$g' \circ f' = g' \circ g_1 \circ f = g \circ f. \qquad \square$$

Theorem 3. *Let L_1, L_2, L_3 be three nonconcurrent lines, and let $f: L_1 \overset{u}{\barwedge} L_2$ and $g: L_2 \overset{v}{\barwedge} L_3$ be perspectivities. If L_2' is a third line on $L_2 \cap L_3$, but not on u, then there exist perspectivities $f' : L_1 \overset{u}{\barwedge} L_2'$ and $g' : L_2' \overset{w}{\barwedge} L_3$, with w on uv, such that $g' \circ f' = g \circ f$.*

The proof of Theorem 3 is so similar to that of Theorem 2 that we omit it. We are now ready for the main theorem.

Theorem 4. *A projectivity between distinct ranges may be expressed as the composition of at most two perspectivities.*

PROOF. We shall show that any sequence of three perspectivities between distinct ranges can be reduced to two perspectivities. To that end, let

$$f: L_1 \overset{u}{\barwedge} L_2 \overset{v}{\barwedge} L_3 \overset{w}{\barwedge} L_4$$

be a chain of three perspectivities, with $L_1 \neq L_4$.

Case 1. If L_1, L_2, L_3 are concurrent, then by Theorem 1 there is a point x on uv such that

$$f: L_1 \overset{x}{\barwedge} L_3 \overset{w}{\barwedge} L_4.$$

Case 2. If L_2, L_3, L_4 are concurrent, then by Theorem 1 there is a point y on vw such that

$$f: L_1 \overset{u}{\barwedge} L_2 \overset{y}{\barwedge} L_4.$$

Case 3. If no three of L_1, L_2, L_3, L_4 are concurrent, let L_0 be the line on $L_1 \cap L_2$ and $L_3 \cap L_4$ (see Figure 3.13). If L_0 is not on v, then we may apply Theorem 2 to obtain

$$f: L_1 \overset{x}{\barwedge} L_0 \overset{v}{\barwedge} L_3 \overset{w}{\barwedge} L_4,$$

and then apply Case 2. If L_0 is on v but not on u, then we may apply Theorem 3 to obtain

$$f: L_1 \overset{u}{\barwedge} L_0 \overset{x}{\barwedge} L_3 \overset{w}{\barwedge} L_4,$$

Figure 3.13

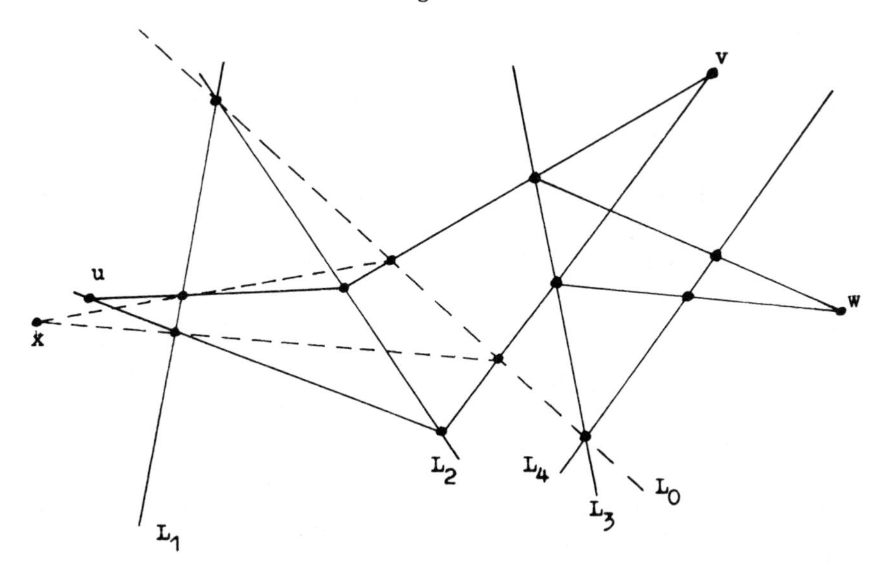

and then apply Case 2. If L_0 is on both u and v but not on w, then we may apply Theorem 2 to get

$$f: L_1 \overset{u}{\barwedge} L_2 \overset{v}{\barwedge} L_0 \overset{x}{\barwedge} L_4,$$

and then apply Case 1. If u, v, and w are all on L_0, let L_3' be another line on $L_3 \cap L_4$, not on v. Then f becomes

$$f: L_1 \overset{u}{\barwedge} L_2 \overset{v}{\barwedge} L_3' \overset{x}{\barwedge} L_4$$

by Theorem 3. Then let L_0' be on $L_1 \cap L_2$ and $L_3' \cap L_4$. Since L_0' is not on v, we now apply Theorem 2 to get

$$f: L_1 \overset{y}{\barwedge} L_0' \overset{v}{\barwedge} L_3' \overset{x}{\barwedge} L_4,$$

and then apply Case 2.

Case 4. If L_1, L_3, L_4 are concurrent, let L_3' be a third line on $L_2 \cap L_3$, not on w; then by Theorem 2, f becomes

$$f: L_1 \overset{u}{\barwedge} L_2 \overset{x}{\barwedge} L_3' \overset{w}{\barwedge} L_4.$$

and we may apply Case 3.

Case 5. If L_1, L_2, L_4 are concurrent, let L_2' be another line on $L_2 \cap L_3$, not on u; then by Theorem 3, f becomes

$$f: L_1 \overset{u}{\barwedge} L_2' \overset{x}{\barwedge} L_3 \overset{w}{\barwedge} L_4.$$

Now we apply Case 3. □

Corollary. *A projectivity from a range to itself may be expressed as the composition of at most three perspectivities.*

The proof of the corollary is left as an exercise. Thus, in a Desarguesian plane, any projectivity can be expressed as a composition of at most three perspectivities.

A second objective in this section is to investigate the existence of certain collineations in a Desarguesian plane. We complete this section with two theorems on this topic, which also are useful later.

Theorem 5. *If c, p, and p' are three collinear points in the Desarguesian plane π and A is a line not on p or p', then there is a central collineation on π with center c and axis A that carries p to p'.*

PROOF. We define a function $f: \pi \to \pi$ as follows: If x is a point of A, let $f(x) = x$. If x is not on A and not on cp, let $q = px \cap A$ and let $f(x) = p'q \cap cx$ (see Figure 3.14). If x is on cp, let y be any point not on cp or A; then let $t = xy \cap A$ and $f(x) = tf(y) \cap cp$ (see Figure 3.15). We now show that f is the required central collineation.

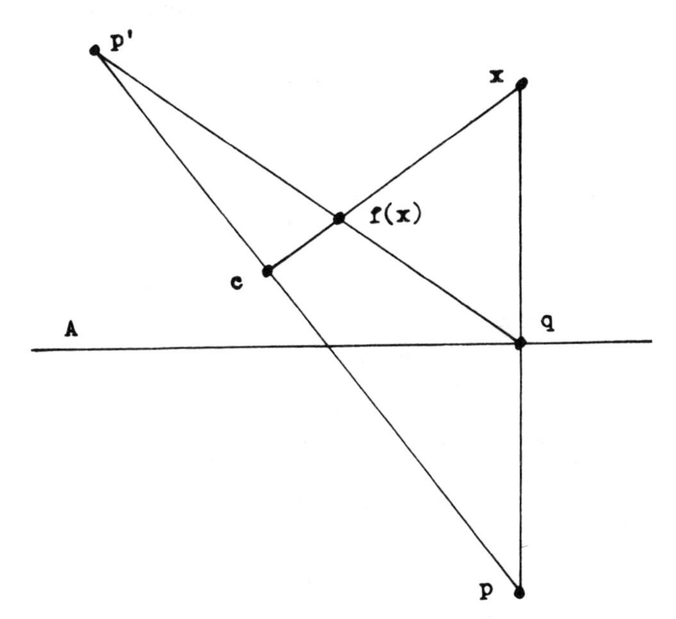

Figure 3.14

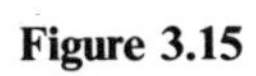

Figure 3.15

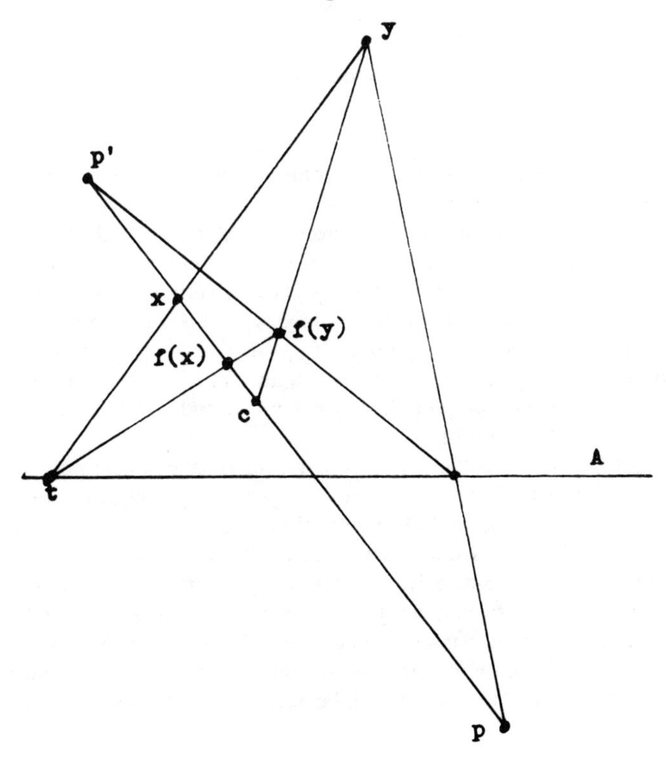

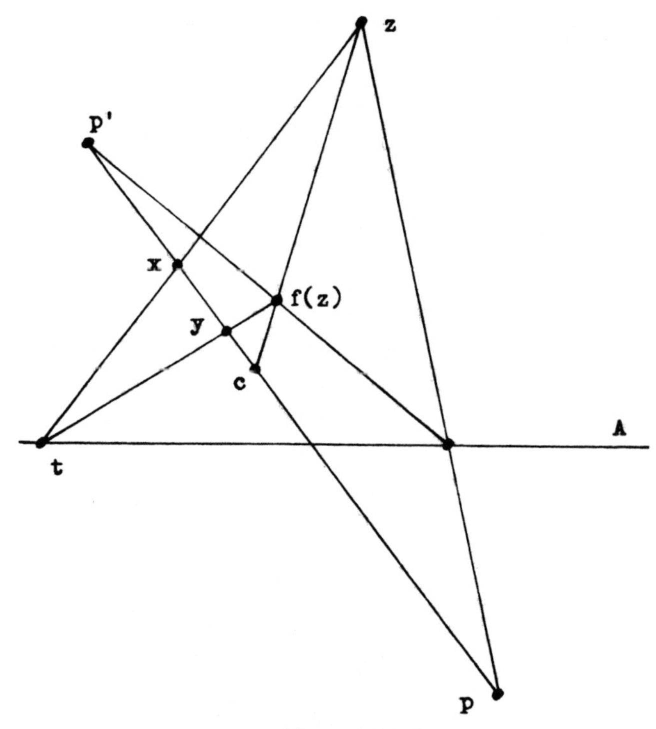

Figure 3.16

It is easily verified that $f(c) = c$; also, f holds A pointwise fixed. Thus we need only show that f is a collineation.

To show that f is one-to-one, suppose $f(x) = f(y)$. If $y \in A$, then $f(x) = y$, so $x \in A$ and $x = y$. Similarly, if $x \in A$, then $x = y$. If y is not on cp, then $f(y)$ is not on cp; hence, if $x \in cp$, then $y \in cp$, and $x = y$ is easily verified. Finally, if neither x nor y is on A or cp, then $x = y$ is easily obtained. Hence f is one-to-one.

To show that f is onto, let y be any point. If $y \in A$, then $f(y) = y$. If y is on cp, let z be any point not on cp or A, let $t = yf(z) \cap A$, and let $x = tz \cap cp$. Then $f(x) = y$ (see Figure 3.16). If y is not on cp or A, let $q = p'y \cap A$ and $x = pq \cap cy$ (see Figure 3.17); then $f(x) = y$. Hence f is onto.

To show that f preserves collinearity, let x, y, z be three collinear points. If none of x, y, or z is on cp and p is not on xy, then $\triangle pxy$ and $\triangle p'f(x)f(y)$ form a central couple with center c (see Figure 3.18). Hence they also form an axial couple. Since $px \cap p'f(x)$ and $py \cap p'f(y)$ are on A, the axis of the couple is A. Let $xy \cap A = t$; then also $f(x)f(y) \cap A = t$. Similarly, $\triangle pxz$ and $\triangle p'f(x)f(z)$ form a central, hence axial, couple with axis A, so that $f(x)f(z) \cap A = t$. Hence $f(x)$, $f(y)$, and $f(z)$ are collinear. If none of x, y, or z is on cp and p does lie on xy, then let $xy \cap A = q$. Then $f(x)$, $f(y)$, $f(z)$ all lie on $p'q$, and so are collinear. If two of x, y, and z lie on cp, so does the third, and $f(x)$, $f(y)$, $f(z)$ all lie on cp. If exactly one, say z, lies on cp,

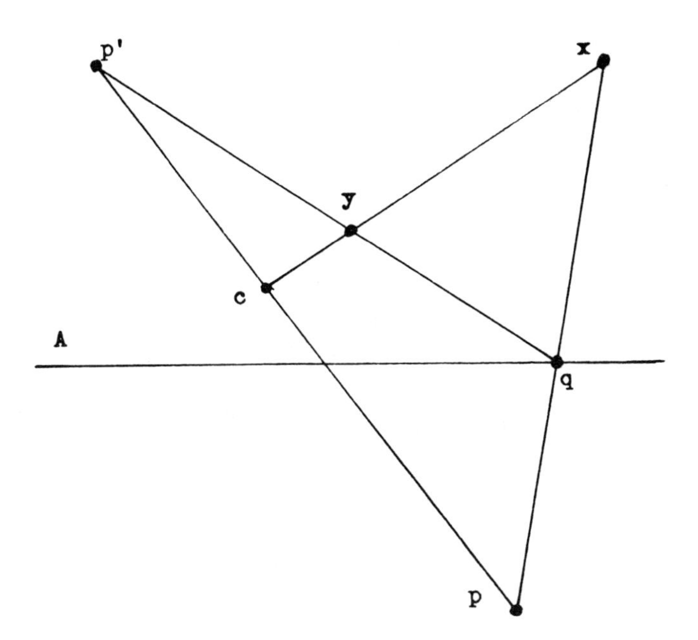

Figure 3.17

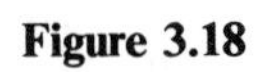
Figure 3.18

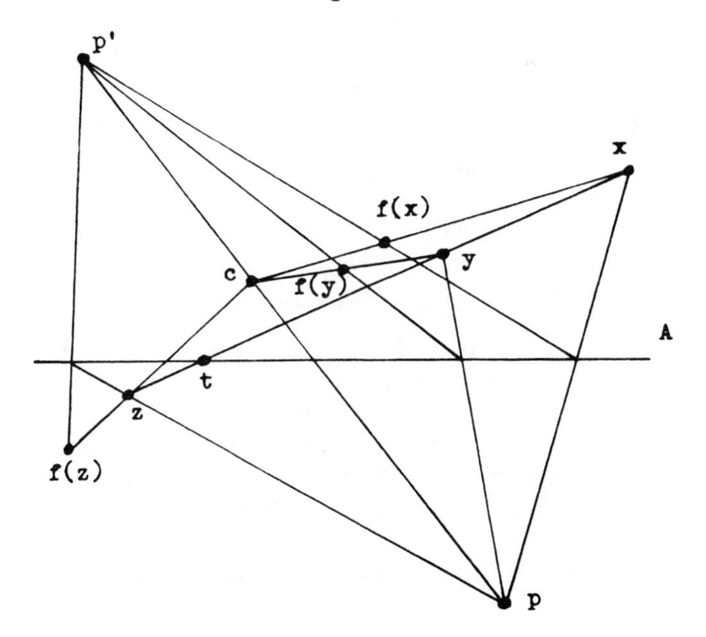

let $t = xy \cap A$. Then $f(x)$, $f(y)$, and $t = f(t)$ are collinear, and $f(z) = cp \cap f(x)t$ is collinear with $f(x)$ and $f(y)$. Thus f preserves collinearity. Hence f is the required collineation. □

Theorem 6. *If $pqrs$ and $p'q'r's'$ are four-points in a Desarguesian plane π, then there is a projective collineation on π carrying p to p', q to q', r to r', and s to s'.*

PROOF. Let A_1 be an arbitrary line not on p or p'. Let f_1 be the elation on π with axis A_1 that carries p to p' (f_1 exists by Theorem 5). Let $q_1 = f_1(q)$, $r_1 = f_1(r)$, $s_1 = f_1(s)$. Let A_2 be a line on p' but not on q_1 or q'. Let f_2 be the elation on π with axis A_2 that carries q_1 to q'. Let $r_2 = f_2(r_1)$, $s_2 = f_2(s_1)$. Since $f_2 \circ f_1$ is a collineation, r_2 is not on $p'q'$. Let f_3 be the elation on π with axis $p'q'$ that carries r_2 to r'. Let $s_3 = f_3(s_2)$, and let $s_4 = q's_3 \cap p's'$. Since $f_3 \circ f_2 \circ f_1$ is a collineation, s_3 is not on $p'r'$. Also, s_4 is not on $p'r'$, since it is on $p's'$. Let f_4 be the homology on π with axis $p'r'$ and center q' that carries s_3 to s_4. Now s_3 is not on $q'r'$, so s_4 is not on $q'r'$. Neither is s' on $q'r'$. Let f_5 be the homology on π with axis $q'r'$ and center p' that carries s_4 to s'. Then

$$f = f_5 \circ f_4 \circ f_3 \circ f_2 \circ f_1$$

is the required projective collineation. □

Exercises 3.2

1. Prove the corollary to Theorem 4.

2. Let $f\colon L \sim L$ be a projectivity in a Desarguesian plane for which a fixed point is known. Can f be expressed as the composition of two perspectives?

3. Let π be a Desarguesian plane satisfying Fano's axiom, let L be a line in π, and let p, q, r be three points on L. Express the projectivity $f\colon L(pqr) \sim L(prq)$ as a composition of two perspectivities.

4. With f as in Exercise 3, find a point $s \neq p$ such that $f(s) = s$.

5. Is f in Exercise 3 an involution?

Section 3.3. Coordinates in Desarguesian Planes

In a Desarguesian plane, it is possible to construct an algebraic structure out of points, which may then be used to introduce coordinates in the plane. This process will be carried out in this section.

We shall begin with a Desarguesian plane π and shall let L be a line in π. Let o, u, i be three points on L. We shall denote the set of points on L distinct from u by $L(oui)$. That is, $L(oui) = L \backslash \{u\}$. We call $L(oui)$ the *open set* on L with *origin* o, *unit point* i, and *ideal point* u.

We now define operations of addition $(+)$ and multiplication $(\cdot)$ on the open set $L(oui)$.

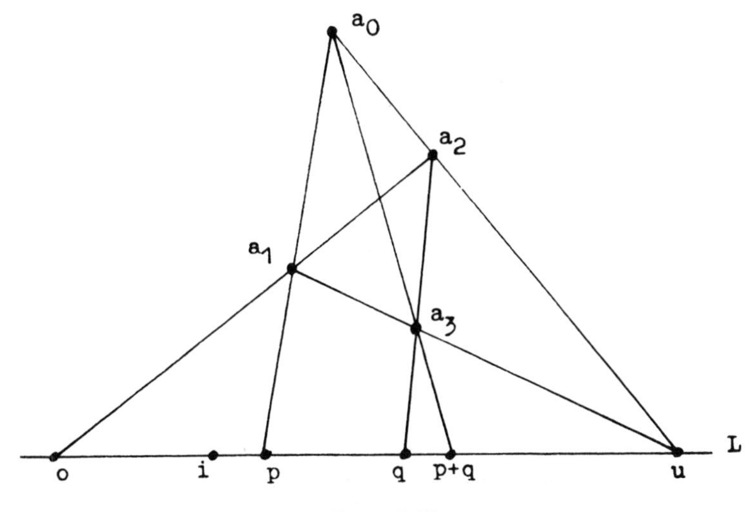

Figure 3.19

Let $p, q \in L(oui)$. Let a_0 be a point not on L, and let a_1 be a third point on a_0p. Let $a_2 = a_1o \cap a_0u$ and $a_3 = a_2q \cap a_1u$. We then define the *sum* of p and q to be the point (see Figure 3.19)

$$p + q = a_0a_3 \cap L.$$

Let $p, q \in L(oui)$. Let b_0 be a point not on L, and let b_1 be a third point on b_0. Let $b_2 = b_1i \cap b_0u$ and $b_3 = b_2q \cap b_1o$. We then define the *product* of p and q to be the point (see Figure 3.20)

$$p \cdot q = b_0b_3 \cap L.$$

Theorem 1. *The operations* $+$ *and* $\cdot$ *on* $L(oui)$ *are well defined, and* $L(oui), +, \cdot$ *is a division ring.*

PROOF. To show that addition is well defined, note that $p \neq u$ and $q \neq u$. If neither $p = o$ nor $q = o$, then $a_0a_1a_2a_3$ can be shown to be a four-point. Hence

$$\{(u, u), (p, q), (o, p + q)\}$$

is a quadrangular set, so $p + q$ is unique by Theorem 3.1.6. If $p = o$ or $q = o$, you can easily verify that $o + q = q$ and $p + o = p$. Hence $+$ is well defined. Note also that $p + q \neq u$, so $L(oui)$ is closed under $+$. Similarly, for multiplication if $p \neq i$ and $q \neq i$, then

$$\{(o, u), (p, q), (i, p \cdot q)\}$$

is a quadrangular set. If $p = i$ or $q = i$, it is easily seen that $i \cdot q = q$ and $p \cdot i = p$. Thus $\cdot$ is well defined. Also, $L(oui)$ is closed under $\cdot$.

We have already noted that o is an additive identity and i is a multiplicative identity. For any point p, the additive inverse of p is the harmonic conjugate of p

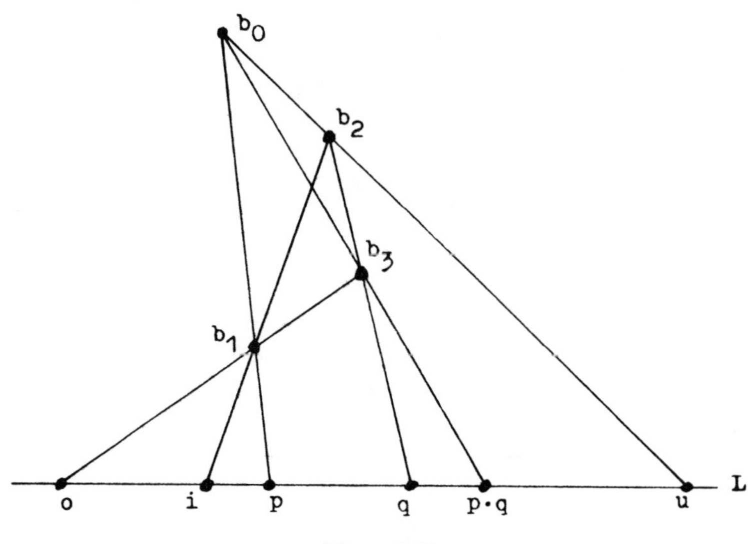

Figure 3.20

with respect to o and u (see Figure 3.21), and for any point $p \neq o$, a multiplicative inverse exists (see Figure 3.22).

It remains to be shown that $+$ and $\cdot$ are associative, $+$ is commutative, and $\cdot$ distributes over $+$. Given points $p, q, r \in L(\dot{o}ui)$, construct $p + q$ using four-point $a_0a_1a_2a_3$ (see Figure 3.23), and construct $(p + q) + r$ using four-point $a_0a_3a_4a_5$. Then construct $q + r$ using four-point $a_2a_3a_4a_5$, and then construct $p + (q + r)$ using four-point $a_0a_1a_2a_5$. Clearly $p + (q + r)$ is the same point as $(p + q) + r$. Hence $+$ is associative. In an identical manner, it can be shown that $\cdot$ is associative (see Figure 3.24). To show that $+$ is commutative and to show that $\cdot$ distributes over $+$ are left as exercises. □

Figure 3.21

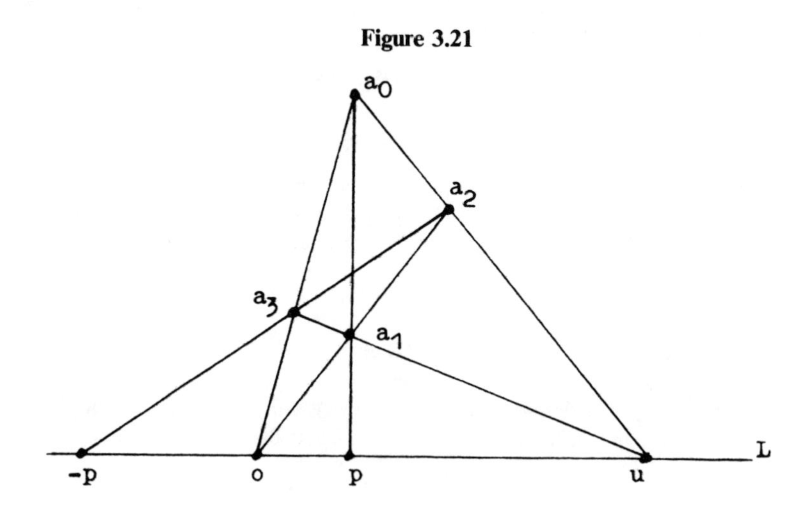

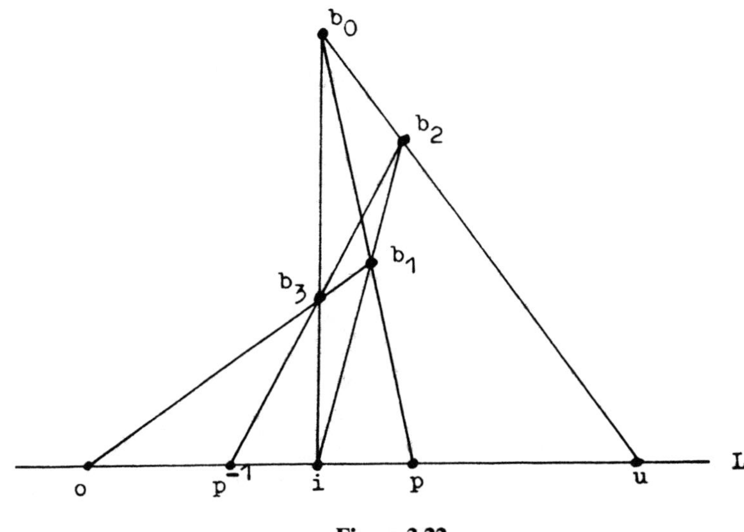

Figure 3.22

Thus it is shown that $L(oui), +, \cdot$ is a division ring. It cannot be shown that $\cdot$ is commutative, however, without assuming a stronger property of the plane, in general.

We now have the curious fact that out of strictly geometric ideas we obtain an algebraic structure.

Figure 3.23

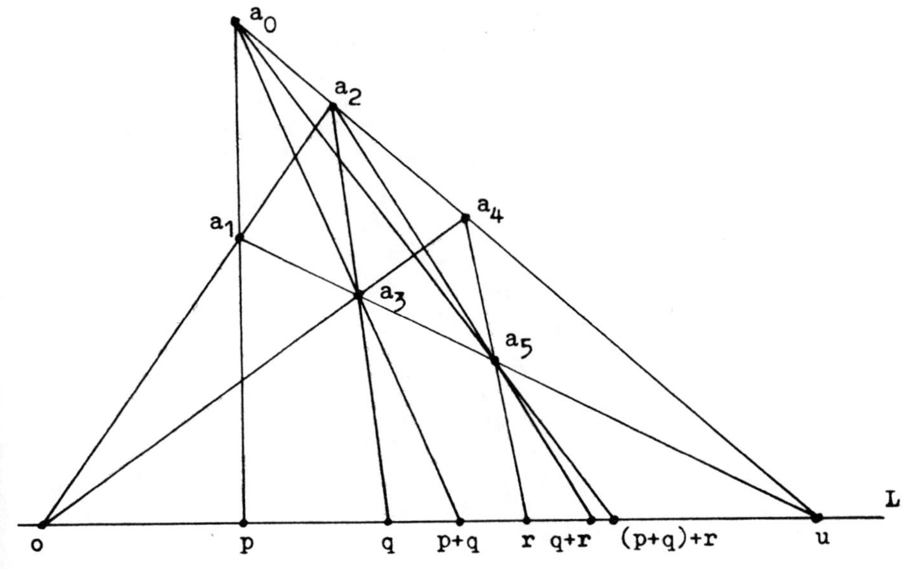

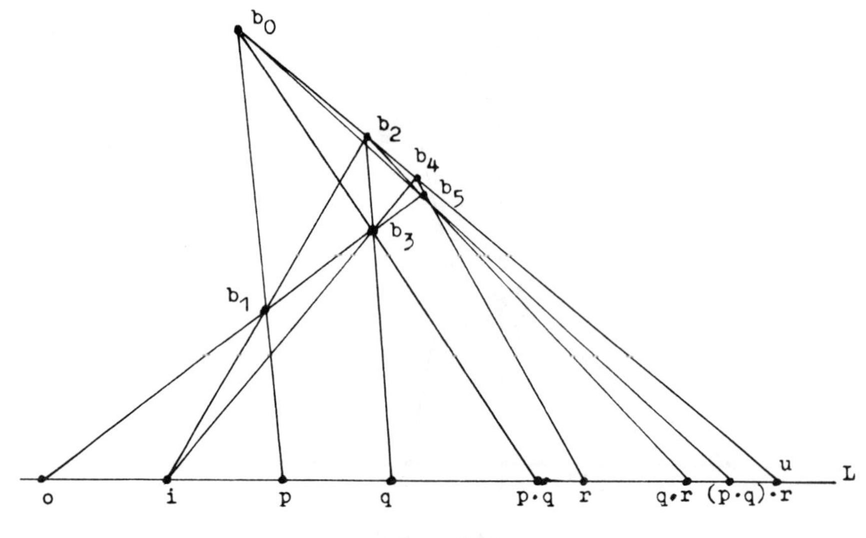

Figure 3.24

The next theorem shows that we get only one algebraic structure in any plane in this manner.

Theorem 2. *If L and L' are lines in a Desarguesian plane π, and o, u, i are three points on L, and o', u', i' are three points on L', then division rings $L(oui)$ and $L'(o'u'i')$ are isomorphic.*

PROOF. Let v be a point not on L, and let e be a third point on vi. Then $uvoe$ is a four-point. Similarly, let v' be a point not on L', and let e' be a third point on $v'i'$. Then $u'v'o'e'$ is a four-point. Let f be a projective collineation on π that carries u to u', v to v', o to o', and e to e'; f exists by Theorem 3.2.6. Then $f(i) = i'$. Now let $p, q \in L(oui)$. Let $a_0 = v$, $a_1 = vp \cap oe$, $a_2 = oe \cap uv$, and $a_3 = a_2q \cap a_1u$. Then $p + q = a_0a_3 \cap L$. Let $f(p) = p'$, $f(q) = q'$. Then $f(a_1) = v'p' \cap o'e'$; let $f(a_1) = a_1'$. Also $f(a_2) = o'e' \cap u'v'$; let $f(a_2) = a_2'$. Moreover, $f(a_3) = a_2'q' \cap a_1'u'$; let $f(a_3) = a_3'$. Hence $f(p + q) = f(a_0a_3 \cap L) = v'a_3' \cap L' = p' + q' = f(p) + f(q)$; thus f is an additive homomorphism. Similarly, $f(p \cdot q) = f(p) \cdot f(q)$, so f is a multiplicative homomorphism. Since $f(L) = L'$ and f is a bijection,

$$f|_{L(oui)} : L(oui) \to L'(o'u'i')$$

is an isomorphism of division rings. □

Henceforth, we shall denote by D_π the division ring isomorphic to any open set on any line in π. That is, associated with any Desarguesian plane π is a division ring D_π.

The obvious question now is whether π is related to the plane π_{D_π}. The answer is yes.

Theorem 3. *If π is a Desarguesian plane, then π is isomorphic to π_{D_π}.*

PROOF. Let $uvoe$ be any four-point in π. Let $L = ou$, let $i = ve \cap L$, and let $D_\pi = L(oui)$. We now define

$$f : \pi \to \pi_{D_\pi}$$

as follows:

First, set $f(u) = [i,o,o]$, $f(v) = [o,i,o]$, $f(o) = [o,o,i]$, $f(e) = [i,i,i]$. If $p_1 \in L(oui)$, that is, if $p_1 \in L \setminus \{u\}$, set $f(p_1) = [p_1, o, i]$.

Before continuing to define f, let $L' = ov$ and $i' = ue \cap L'$. Then $L'(voi')$ is isomorphic to D; for each $a \in D_\pi$, let a' be the element of $L'(voi')$ corresponding to a under that isomorphism.

Now returning to f, if $p_2' \in L' \setminus \{v\}$, set $f(p_2') = [o, p_2, i]$. If p is a point not on uv, let $p_1 = vp \cap L$ and $p_2' = up \cap L'$, and set $f(p) = [p_1, p_2, i]$. Finally, if $p_3 \in uv$, let p be a third point on op_3; if $f(p) = [p_1, p_2, i]$, set $f(p_3) = [p_1, p_2, o]$. Note that at each step, the definition of f is consistent with the preceding steps.

It now must be shown that f is bijective and preserves collinearity. To show that f is onto, let $[a,b,c]$ be any point of π_{D_π}. If $c \neq o$, let $p_1 = c^{-1} \cdot a$, $p_2 = c^{-1} \cdot b$, and $p = p_1 v \cap p_2' u$. Then $f(p) = [p_1, p_2, i] = [c^{-1} \cdot a, c^{-1} \cdot b, i] = [a,b,c]$. If $c = o$, let $p = av \cap ub'$ and $p_3 = op \cap uv$. Then $f(p) = [a,b,i]$, so $f(p_3) = [a,b,o] = [a,b,c]$. Hence f is onto. To show that f is one-to-one, suppose $f(p) = f(q) = [a,b,c]$. If $c \neq o$, let $p_1 = c^{-1} \cdot a$, $p_2 = c^{-1} \cdot b$. Then $[a,b,c] = [p_1, p_2, i]$, so $p = q = vp_1 \cap up_2'$. If $c = o$, let $r = va \cap ub'$; then $p = q = or \cap uv$. Hence f is one-to-one.

To show that f preserves collinearity, we consider several cases.

Case 1. Any point on line uv has image under f of the form $[p_1, p_2, o]$, which always lies on line $\langle o,o,i \rangle$. Hence, points collinear on uv have images collinear on $\langle o,o,i \rangle$.

Case 2. Let M be a line on o. Let $p_3 = M \cap uv$, and let $f(p_3) = [p_1, p_2, o]$. Then any point on M has image of the form $[k \cdot p_1, k \cdot p_2, i]$, which is on line $\langle p_1^{-1}, -p_2^{-1}, o \rangle$. Hence, points collinear on M have images collinear on $\langle p_1^{-1}, -p_2^{-1}, o \rangle$.

Case 3. Let M be a line on u, but not on v. Let $p_2' = M \cap L'$; then $f(p_2') = [o, p_2, i]$. Then any point on M has image of the form $[p_1, p_2, i]$, which is on line $\langle o, i, -p_2 \rangle$. Hence, points collinear on M have images collinear on $\langle o, i, -p_2 \rangle$.

Case 4. Let M be a line on v, but not on u. Let $p_1 = M \cap L$; then $f(p_1) = [p_1, o, i]$. Then any point on M has image of the form $[p_1, p_2, i]$, which is on line $\langle i, o, -p_1 \rangle$. Hence, points collinear on M have images collinear on $\langle i, o, -p_1 \rangle$.

Case 5. Let M be a line not on o, u, or v. Let $q_1 = M \cap L$, $q_2' = M \cap L'$, and $q_3 = M \cap uv$. If $q = oq_3 \cap uq_2'$, then $vq \cap L = -q_1$, for $(ou, q_1(-q_1))$

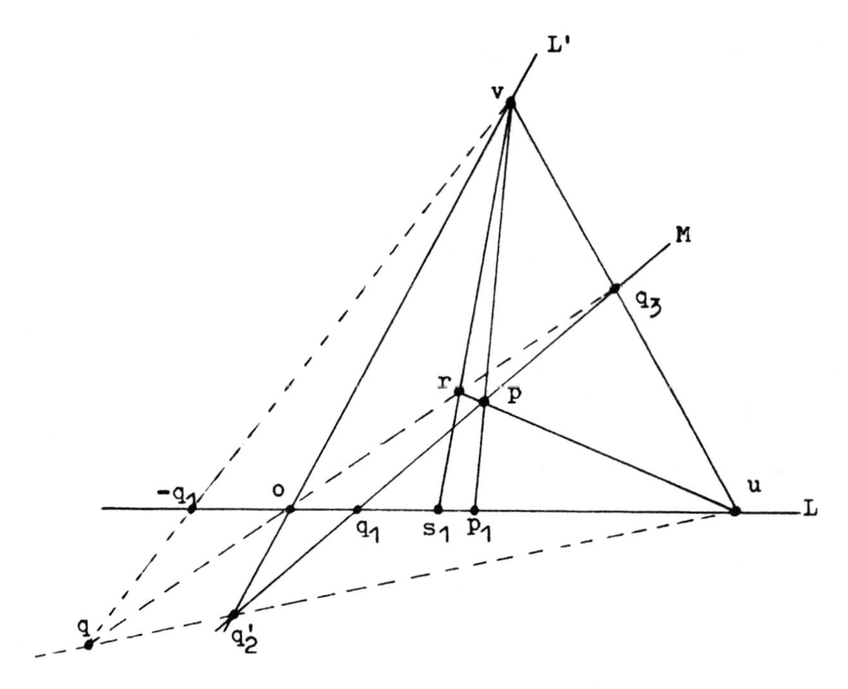

Figure 3.25

is a harmonic set (see Figure 3.25). Thus $f(q) = [-q_1, q_2, i]$, so $f(q_3) = [-q_1, q_2, o]$. Now let p be any point on M. Let $p_1 = up \cap L$, and let $r = up \cap op_3$. Let $s_1 = vr \cap L$. Forming the sum of s_1 and q_1, using the four-point urq_3p, we find $s_1 + q_1 = p_1$. Hence $s_1 = p_1 - q_1$, so $f(p) = [p_1, p_2, i]$ implies $f(r) = [p_1 - q_1, p_2, i]$. Thus $f(q_3) = [p_1 - q_1, p_2, o]$ as well. Hence

$$[p_1 - q_1, p_2, o] = [-q_1, q_2, o],$$

so that

$$p_1 = (i - k) \cdot q_1, \qquad p_2 = k \cdot q_2.$$

Hence $f(p) = [(i-k) \cdot q_1, k \cdot q_2, i]$, which is on line $\langle q_1^{-1} \cdot q_2, i, -q_2 \rangle$. Thus points collinear on M have images collinear on $\langle q_1^{-1} \cdot q_2, i, -q_2 \rangle$. This completes the proof. □

The isomorphism

$$f : \pi \to \pi_{D_\pi}$$

of Theorem 3 is called a *coordinatization* of π, for it assigns to each point p of π a point $[a,b,c]$ of π_{D_π}, called a *set of coordinates* for p. The four-point $uvoe$ is called the *reference quadrangle* of the coordinatization. The division ring D_π is also called the *coordinate ring* of π, and π is said to be *coordinatized over* D_π by f. We can therefore restate Theorem 3 as follows:

Theorem 4. *Any Desarguesian plane π can be coordinatized over D_π, using any four-point as reference quadrangle.*

From another point of view, Theorem 3 tells us that if we start with a Desarguesian plane π, we can obtain a division ring D_π, from which in turn we can obtain another Desarguesian plane π_{D_π}; but the latter is isomorphic to the former. In a similar manner, if we start with a division ring D, we can obtain a Desarguesian plane π_D, from which in turn we can obtain a division ring D_{π_D}; as you may suspect, the latter is again isomorphic to the former.

Theorem 5. *If D is a division ring, then D is isomorphic to D_{π_D}.*

PROOF. In π_D, let L be the line $\langle 0,0,1\rangle$, and let $o=[0,0,1]$, $u=[0,1,0]$, and $i=[1,1,0]$. Then $L(oui)$ is isomorphic to D_{π_D}. But by the definitions of $+$ and $\cdot$, we find that

$$[1,a,0]+[1,b,0]=[1,a+b,0],$$

$$[1,a,0]\cdot[1,b,0]=[1,ab,0].$$

Hence $f: D\to L(oui)$, defined by $f(a)=[1,a,0]$, is an isomorphism. Hence D is isomorphic to D_{π_D}. □

Exercises 3.3

1. Prove that $+$ is commutative.

2. Prove that $\cdot$ distributes over $+$.

3. In the proof of Theorem 5, verify that $[1,a,0]+[1,b,0]=[1,a+b,0]$ and $[1,a,0]\cdot[1,b,0]=[1,ab,0]$.

4. Explain why a projective collineation on a Desarguesian plane is sometimes called a *change of coordinates*.

Section 3.4. Pappian Planes

In the fourth century, the Greek geometer Pappus of Alexandria proved a theorem in Euclidean geometry that bears his name today. A projective plane that satisfies the projective version of the theorem of Pappus is called a *Pappian plane*.

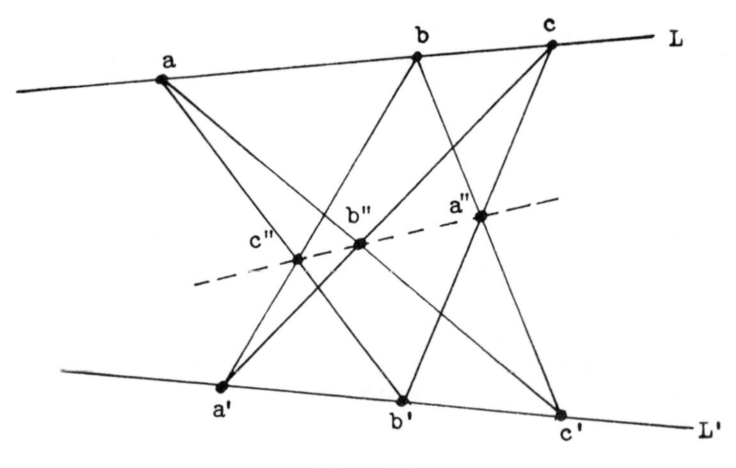

Figure 3.26

The Theorem of Pappus. *If a, b, c are three points on a line L and a', b', c' are three points on a second line L', none of these points being on both L and L', then the points $a'' = bc' \cap b'c$, $b'' = ca' \cap c'a$, $c'' = ab' \cap a'b$ are collinear (see Figure 3.26).*

The points a'', b'', c'' in the theorem of Pappus are often called the *cross-joins* of the triples (a,b,c) and (a',b',c'), and the line on a'', b'', c'', is called the *Pappus line* of the triples (a,b,c) and (a',b',c').

We first show that the real projective plane is Pappian, by proving a more general result.

Theorem 1. *If F is a field, then π_F is a Pappian plane.*

PROOF. Let L and L' be two lines in π_F, and let $p = L \cap L'$. Let q be a second point on L, and let q' be a second point on L'. We will now use p and q as base points in a parametrization of L. Since none of the points a, b, c is p, we may give them parameters as follows:

$$a : (\alpha, 1),$$
$$b : (\beta, 1),$$
$$c : (\gamma, 1).$$

Similarly, using p and q' as base points in a parametrization of L', we may give a', b', c' parameters

$$a' : (\alpha', 1),$$
$$b' : (\beta', 1),$$
$$c' : (\gamma', 1).$$

Now since $a'' = bc' \cap b'c$, we can express a'' parametrically in terms either of b and c' or of b' and c. Hence there exist λ, μ', λ', and $\mu \in F$ such that the ith coordinate of a'' is

$$a_i'' = \lambda b_i + \mu' c_i' = \lambda' b_i' + \mu c_i.$$

Hence

$$\lambda(\beta p_i + q_i) + \mu'(\gamma' p_i + q_i') = \lambda'(\beta' p_i + q_i') + \mu(\gamma p_i + q_i),$$

or

$$(\lambda\beta - \mu'\gamma' - \lambda'\beta' - \mu\gamma)p_i + (\lambda - \mu)q_i + (\mu' - \lambda')q_i' = 0.$$

But since p, q, and q' are noncollinear, we must have

$$\begin{aligned} \lambda\beta - \mu'\gamma' - \lambda'\beta' - \mu\gamma &= 0, \\ \lambda - \mu &= 0, \\ \mu' - \lambda' &= 0. \end{aligned}$$

Solving, we find

$$\lambda = \mu = \beta' - \gamma',$$

$$\lambda' = \mu' = \beta - \gamma.$$

Thus

$$\begin{aligned} a_i'' &= \lambda b_i + \mu' c_i' \\ &= (\beta' - \gamma')(\beta p_i + q_i) + (\beta - \gamma)(\gamma' p_i + q_i') \\ &= (\beta\beta' - \gamma\gamma')p_i + (\beta' - \gamma')q_i + (\beta - \gamma)q_i'. \end{aligned}$$

(It is this last step that employs the commutativity of the field F.) In a similar fashion, we find

$$b_i'' = (\gamma\gamma' - \alpha\alpha')p_i + (\gamma' - \alpha')q_i + (\gamma - \alpha)q_i',$$

$$c_i'' = (\alpha\alpha' - \beta\beta')p_i + (\alpha' - \beta')q_i + (\alpha - \beta)q_i'.$$

Hence

$$a_i'' + b_i'' + c_i'' = 0,$$

and a'', b'', c'' are collinear by Exercise 1.3.11. □

The properties of projectivities in a Pappian plane are based upon the following theorem.

Theorem 2. *If L_1, L_2, L_3 are three nonconcurrent lines in a Pappian plane π, and $f: L_1 \barwedge L_2$ and $g: L_2 \barwedge L_3$ are perspectivities such that $g \cdot f: L_1 \sim L_3$ is a projectivity in which point $L_1 \cap L_3$ is its own image, then $g \cdot f$ is a perspectivity.*

PROOF. Let $f: L_1 \overset{u}{\barwedge} L_2$ and $g: L_2 \overset{v}{\barwedge} L_3$. Let $a_1 = L_2 \cap L_3$, $a_2 = L_1 \cap L_3$, and $a_3 = L_1 \cap L_2$ (see Figure 3.27). Let $w = ua_1 \cap va_3$. If $g \cdot f$ carries a_2 (as a point of L_1) to itself (as a point of L_3), then u, v, and a_2 must be collinear. If p_1 is any point on L_1, let $p_2 = f(p_1)$ and $p_3 = g(p_2)$. By the theorem of Pappus, the cross-joins of the

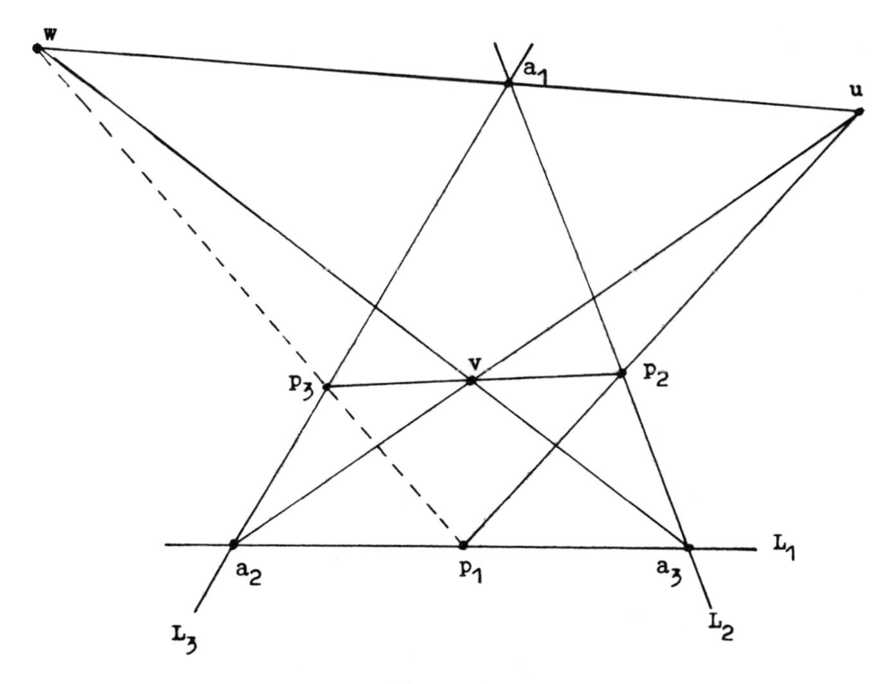

Figure 3.27

triples (u, v, a_2) and (a_3, a_1, p_2) are collinear; hence points

$$ua_1 \cap va_3 = w,$$
$$up_2 \cap a_2a_3 = p_1,$$
$$vp_2 \cap a_2a_1 = p_3$$

are collinear. Hence

$$g \cdot f : L_1 \overset{w}{\barwedge} L_3. \qquad \square$$

With the aid of Theorem 2, we can prove

Theorem 3. *A Pappian plane is Desarguesian.*

PROOF. Let $\triangle abc$ and $\triangle a'b'c'$ form a central couple in Pappian plane π, with center o. Let $L_1 = aa'$, $L_2 = bb'$, $L_3 = cc'$. Let $a'' = bc \cap b'c'$, $b'' = ca \cap c'a'$, $c'' = ab \cap a'b'$ (see Figure 3.28). Consider the projectivity $f: L_1 \overset{c''}{\barwedge} L_2 \overset{a''}{\barwedge} L_3$. Clearly $o = L_1 \cap L_3$ is its own image, but since L_1, L_2, L_3 are concurrent, we may not yet apply Theorem 2. Hence let $q_1 = a''c'' \cap L_1$, $q_2 = a''c'' \cap L_2$, $q_3 = a''c'' \cap L_3$, and let L_1' be a third line on q_1. Let $oa'' \cap L_1' = o'$, $bc \cap L_1' = b_1$, and $b'c' \cap L_1' = b_1'$. Then we may express f as

$$f : L_1(aa'q_1o) \overset{c''}{\barwedge} L_2(bb'q_2o) \overset{a''}{\barwedge} L_1'(b_1b_1'q_1o') \overset{a''}{\barwedge} L_3(cc'q_3o).$$

Now $L_1 \overset{c''}{\barwedge} L_2 \overset{a''}{\barwedge} L_1'$ is a perspectivity by Theorem 2, for q_1 is its own image. Hence $f: L_1 \barwedge L_1' \overset{a''}{\barwedge} L_3$ is also a perspectivity by Theorem 2 for o is its own image. Hence

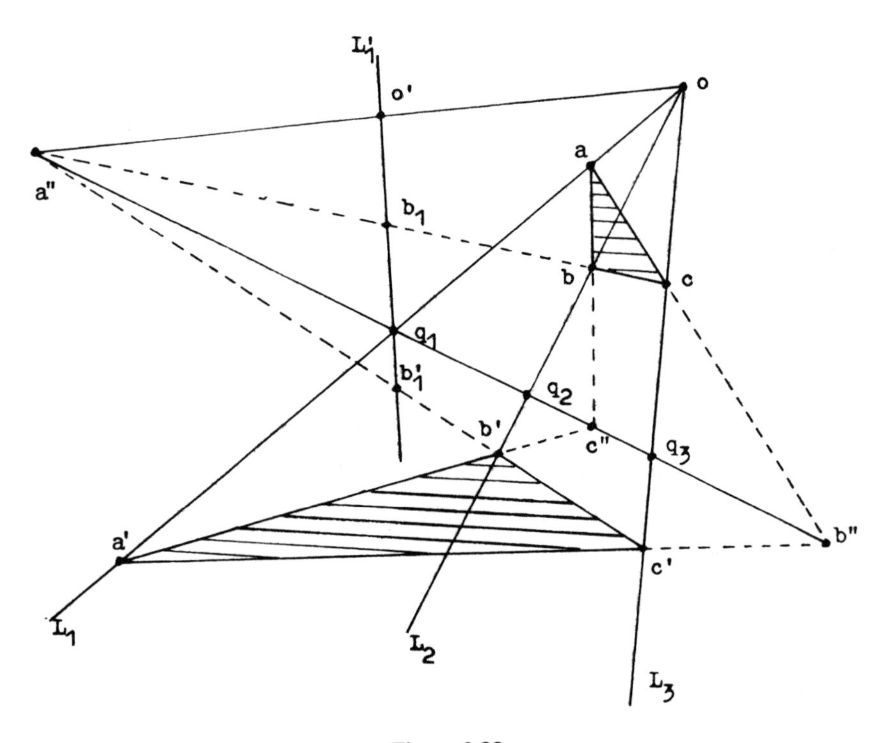

Figure 3.28

$f: L_1(aa'q_1o) \overset{b''}{\barwedge} L_3(cc'q_3o)$, and b'' is on $q_1q_3 = a''c''$. Hence $\triangle abc$ and $\triangle a'b'c'$ form an axial couple, and π is Desarguesian. □

We may now ascribe all the properties of Desarguesian planes to Pappian planes. However, using Theorem 2, we can achieve even stronger results for Pappian planes.

Theorem 4. The Perspectivity Theorem. *If $f: L \sim L'$ is a projectivity between distinct ranges in a Pappian plane and $f(L \cap L') = L \cap L'$, then f is a perspectivity.*

PROOF. By Theorem 3, we may apply Theorem 3.2.4. Hence by Theorem 3.2.1 or by Theorem 2, f must be a perspectivity. □

Theorem 5. The Fundamental Theorem. *In a Pappian plane, a projectivity is uniquely determined by three points and their images.*

PROOF. Let a, b, c be three points on line L, and let a', b', c' be three points on line L'. Suppose there are two projectivities $f_1, f_2: L \sim L'$ carrying a to a', b to b', and c to c'. If $f_1 \neq f_2$, there is some point x on L such that $x' = f_1(x)$ and $x'' = f_2(x)$ are distinct. Now let L_1 and L_2 be two lines, with $p = L_1 \cap L_2$. Let b_1, c_1 be two points of L_1 distinct from p, and let b_2, c_2 be two points of L_2 distinct from p. By Theorem

2.2.1, there exist projectivities

$$g : L_1(pb_1c_1) \sim L(abc),$$
$$h : L'(a'b'c') \sim L_2(pb_2c_2).$$

Let $g^{-1}(x) = y$, $h(x') = y'$, $h(x'') = y''$. Then $y' \neq y''$. Now

$$h \circ f_1 \circ g : L_1(pb_1c_1y) \sim L_2(pb_2c_2y')$$

and

$$h \circ f_2 \circ g : L_1(pb_1c_1y) \sim L_2(pb_2c_2y'')$$

are projectivities carrying p to p, and so must be perspectivities by the perspectivity theorem. By Exercise 2.1.1, $h \circ f_1 \circ g$ and $h \circ f_2 \circ g$ are the same perspectivity, so $y' = y''$, a contradiction. Hence $f_1 = f_2$, and the theorem is proved. □

Theorem 6. *If a projectivity in projective plane π is always uniquely determined by three points and their images, then π is Pappian.*

PROOF. Let L and L' be two lines in π which meet at o, let a, b, c be three points on L, and let a', b', c' be three points on L', all these points distinct from o (see Figure 3.29). Let $a'' = bc' \cap b'c$, $b'' = ca' \cap c'a$, $c'' = ab' \cap a'b$, $p = ab' \cap bc'$, and $q = ac' \cap cb'$. Let $L_1 = ab'$ and $L_2 = ac'$. Consider the projectivity

$$f : L_1(ab'pc'') \overset{b}{\barwedge} L'(ob'c'a') \overset{c}{\barwedge} L_2(aqc'b'').$$

But also

$$L_1(ab'p) \overset{a''}{\barwedge} L_2(aqc');$$

hence, by hypothesis, f is a perspectivity with center a''. Hence a'', b'', c'' are collinear, and π is Pappian. □

Figure 3.29

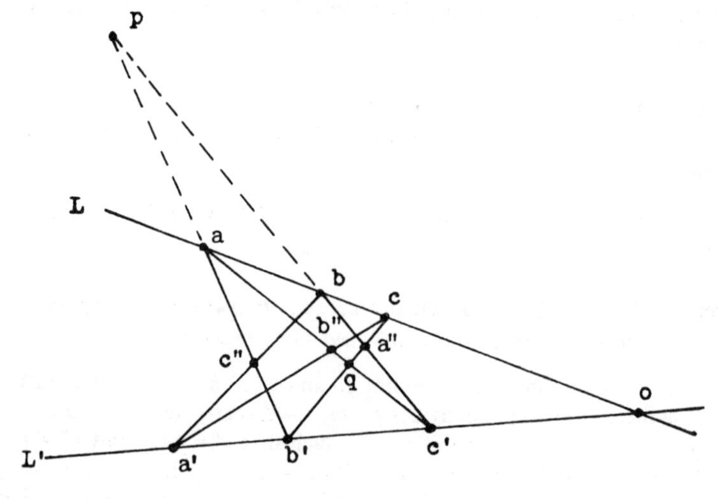

We may combine the fundamental theorem with its converse, Theorem 6, to state that a plane is Pappian if and only if three points and their images uniquely determine a projectivity.

Theorem 2.2.1 gave a method for constructing the image of a point under a projectivity, by resolving the projectivity into a sequence of perspectivities. In a Pappian plane, a simpler method exists, which uses the Pappus line of a triple and its image.

Theorem 7. *Let $f: L(abc) \sim L'(a'b'c')$ be a projectivity between distinct ranges in a Pappian plane. Let P be the Pappus line of the triples (a,b,c) and (a',b',c'). If x is any point on L and $x' = f(x)$, then $ax' \cap a'x$ is on P.*

PROOF. Let $c'' = ab \cap a'b$, $b'' = ac' \cap a'c$, $x'' = a'x \cap P, a_1 = aa' \cap P$, and $x_1 = ax'' \cap L'$ (see Figure 3.30). Let g be the projectivity

$$g : L(abcx) \overset{a'}{\barwedge} P(a, c''b''x'') \overset{a}{\barwedge} L'(a'b'c'x_1).$$

But

$$f : L(abcx) \sim L'(a'b'c'x'),$$

so by the fundamental theorem, $f = g$ and $x' = x_1$. Hence $x'' = ax_1 \cap a'x = ax' \cap a'x$ is on P. $\square$

By a repeated application of Theorem 7, we can show that associated with any projectivity between distinct lines L and L' in a Pappian plane is a line P, called the *axis* of the projectivity, such that if x, y are two points on L and x', y' are their images on L', then $xy' \cap x'y$ is on P. The axis P is just the Pappus line of any three points on L and their images on L'.

In a Pappian plane, we can also show the existence of certain projectivities.

Figure 3.30

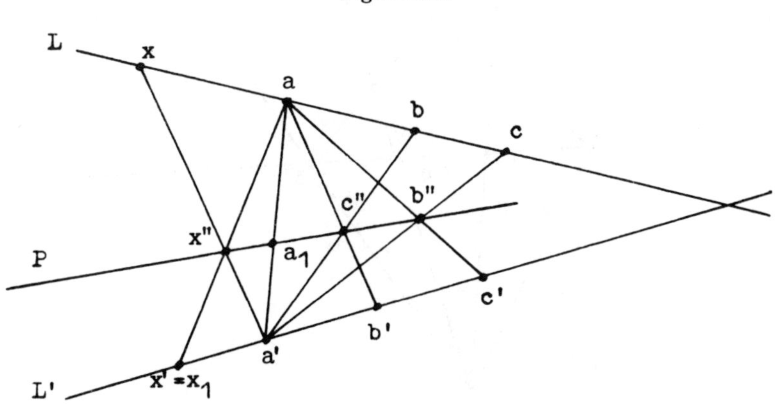

Theorem 8. The Permutation Theorem. *If a, b, c, d are four points on line L in a Pappian plane, then there exist projectivities*

$$f : L(abcd) \sim L(badc),$$
$$g : L(abcd) \sim L(cdab),$$
$$h : L(abcd) \sim L(dcba).$$

PROOF. To show the existence of f, let L' be a line distinct from L, let u be a point not on L or L', and let

$$L(abcd) \overset{u}{\barwedge} L'(a'b'c'd')$$

(see Figure 3.31). Let $v = ca' \cap bd'$. The Pappus line of triples (abc) and $(b'a'd')$ is uv. Let f' be the projectivity $f' : L(abc) \sim L'(b'a'd')$. By Theorem 7, $f'(d) = c'$. Hence, we have

$$f : L(abcd) \sim L'(b'a'd'c') \overset{u}{\barwedge} L(badc).$$

To show the existence of g, let L' and u be as before, and let $w = ad' \cap bc'$ (see Figure 3.32). Then the Pappus line of triples (abc) and $(c'd'a')$ is uw. If g' is the projectivity $g' : L(abc) \sim L'(c'd'a')$, then $g'(d) = b'$ by Theorem 7 again. Hence we have

$$g : L(abcd) \sim L'(c'd'a'b') \overset{u}{\barwedge} L(cdab).$$

The proof of the existence of h is left as an exercise. □

Theorem 9. *Let π be a Desarguesian plane satisfying Fano's axiom. If (ab, cd) is a harmonic set, then there is a projectivity $(abcd) \sim (abdc)$.*

Figure 3.31

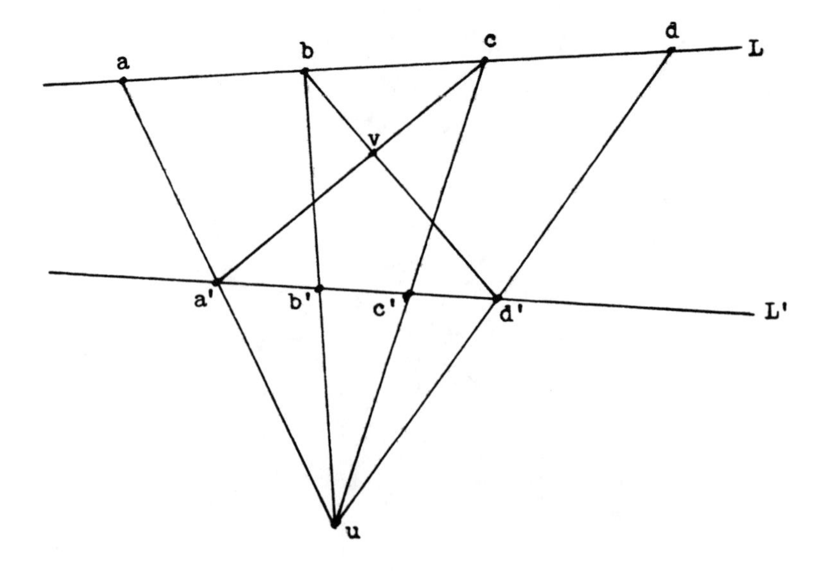

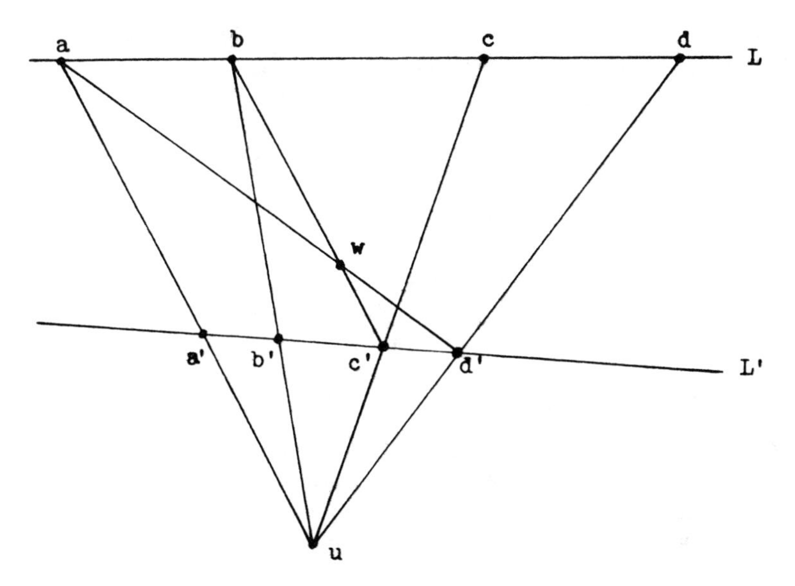

Figure 3.32

PROOF. Let $L = ab$, let p be a point not on L, and let q be a second point on pa. Let $r = qc \cap pb$ and $s = ra \cap qb$; then $d = ps \cap L$ (see Figure 3.33). Let $t = ps \cap qr$, $u = bt \cap rs$, and $v = bt \cap pq$. Then

$$L(abcd)\overset{t}{\barwedge}rs(aurs)\overset{b}{\barwedge}pq(avpq)\overset{t}{\barwedge}L(abdc). \qquad \square$$

Theorem 10. *Let π be a Pappian plane satisfying Fano's axiom. If there is a projectivity $(abcd)\sim(abdc)$, then (ab,cd) is a harmonic set.*

Figure 3.33

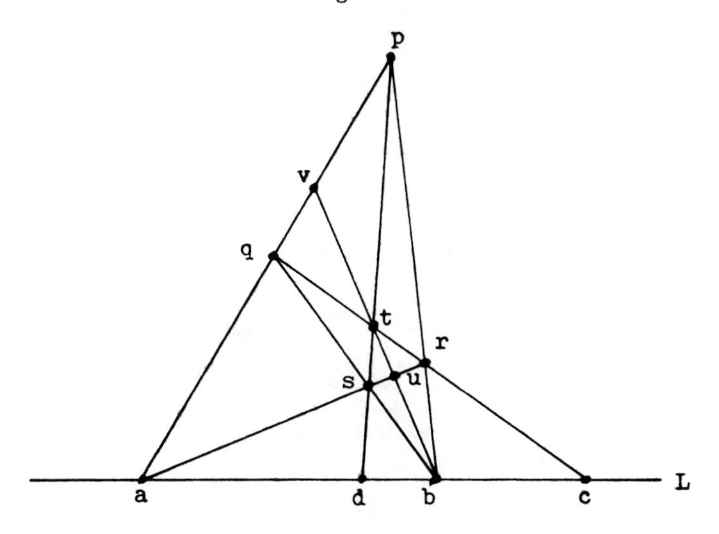

PROOF. If (ab, cd) is not a harmonic set, then neither is (cd, ab), by Exercise 3.1.12. By Theorem 3.1.5, let b' be a point such that (cd, ab') is a harmonic set. Then also (ab', cd) is a harmonic set, so by Theorem 9, there is a projectivity $(ab'cd) \sim (ab'dc)$. But this must be the given projectivity, since it also carries a to a, c to d, and d to c. Hence we have $(abb'cd) \sim (abb'dc)$. But since a, b, and b' are three fixed points, this projectivity must be the identity; it follows that $c = d$, contradicting Fano's axiom (see Exercise 3.1.9). Hence (ab, cd) is a harmonic set. □

Since a Pappian plane is Desarguesian, we can construct a division ring D_π in any Pappian plane π. But since the plane is Pappian, we can prove the commutativity of multiplication in D_π, making D_π a field.

Theorem 11. *If π is a Pappian plane, then D_π is a field.*

PROOF. Given an open set $L(oui)$ in π, let $p, q \in L(oui)$. We construct $p \cdot q$ using the four-point $b_0b_1b_2b_3$ (see Figure 3.34), and then construct $q \cdot p$ using four-point $b_2b_3b_4b_5$. Now the Pappus line of the triples (b_0, p, b_1) and (i, b_3, b_4) is on $p \cdot q$, b_2, and b_5, so $q \cdot p = p \cdot q$. □

By Theorem 3.3.3, we conclude then that a Pappian plane is isomorphic to the plane over a field, or can be coordinatized over a field. We also see that a Desarguesian plane need not be Pappian, for if D is a division ring that is not a field, then π_D is Desarguesian but not Pappian. It is a famous theorem due to Wedderburn (see, for example, Herstein [11, p. 360]) that a finite division ring must be a field. From this fact comes the following geometric theorem, for which a geometric proof has never been given.

Theorem 12. *A finite Desarguesian plane is Pappian.*

The proof is left as an exercise; assume Wedderburn's theorem.

Figure 3.34

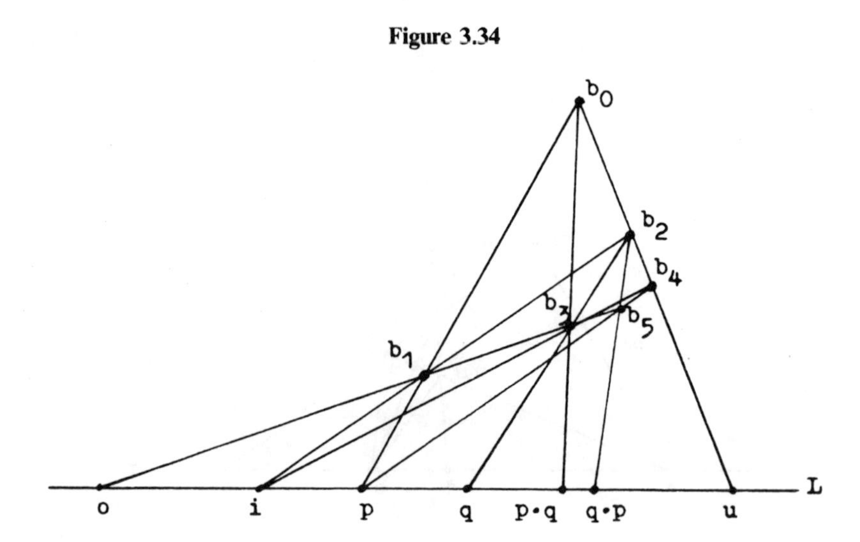

We conclude this section with a discussion of the algebraic consequence of Fano's axiom. First, we must discuss the characteristic of a field.

Let F be a field. If there is a positive integer n such that

$$\underbrace{1 + 1 + \cdots + 1}_{n \text{ times}} = 0$$

in F, we say F has *nonzero characteristic*, and we define the *characteristic* of F, denoted by char(F), to be the smallest such n. If no such positive integer exists, we say F has *zero characteristic* and set char(F) = 0.

Theorem 13. *A Pappian plane π is a Fano plane if and only if* $\text{char}(D_\pi) = 2$.

PROOF. By the definition of addition in $L(oui)$ (Section 3.3), $(ui, o(i + i))$ is a harmonic set. Hence, by Exercise 3.1.9, π is a Fano plane if and only if $i + i = o$, or $\text{char}(D_\pi) = 2$.

On the other hand, then, a Pappian plane satisfies Fano's axiom if and only if $\text{char}(D_\pi) \neq 2$.

Exercises 3.4

*1. Prove that the principle of duality holds in the class of Pappian planes.

2. Prove that the principle of duality holds in the class of Pappian planes satisfying Fano's axiom.

3. Prove that h exists in Theorem 8.

*4. Let π be a Pappian plane satisfying Fano's axiom, and let (ab, cd) be a harmonic set in π. If there is a projectivity $(abcd) \sim (a'b'c'd')$, prove that $(a'b', c'd')$ is a harmonic set.

*5. State and prove the converse of Problem 4.

*6. Prove that if π is a Desarguesian plane and D_π is a field, then π is Pappian.

7. Prove that a Pappian plane is self-dual.

8. Prove Theorem 12.

9. Given two lines L and L' whose point of intersection is inaccessible, and a point p not on L or L', construct the line on p that is concurrent with L and L'.

*10. If $f: L(abc \cdots) \sim L(bac' \cdots)$ is a projectivity in a Pappian plane that interchanges points a and b, prove that f is an involution.

*11. Show that if $\{(a, a'), (b, b'), (c, c')\}$ is a quadrangular set, then the projectivity $(abc) \sim (a'b'c')$ is an involution.

*12. If a and b are fixed points and c and c' are a pair of mates under an involution, prove that (ab, cc') is a harmonic.

Section 3.5. Cross Ratio in π_F

In this section, we restrict our attention to Pappian planes. Because of Theorem 3.4.11, a Pappian plane may be regarded as the plane over a field. Hence, coordinate methods may be used.

Our particular objective in this section is to study a certain invariant called the cross ratio.

Let p_1, p_2, p_3, p_4 be four collinear points in π_F. Relative to some parametrization of the line on which these points lie, let p_i have parameters (λ_i, μ_i) $(i = 1,2,3,4)$. The *cross ratio* of p_1, p_2, p_3, p_4 is the expression

$$R = (p_1p_2p_3p_4) = \frac{\begin{vmatrix} \lambda_1 & \mu_1 \\ \lambda_3 & \mu_3 \end{vmatrix}\begin{vmatrix} \lambda_2 & \mu_2 \\ \lambda_4 & \mu_4 \end{vmatrix}}{\begin{vmatrix} \lambda_1 & \mu_1 \\ \lambda_4 & \mu_4 \end{vmatrix}\begin{vmatrix} \lambda_2 & \mu_2 \\ \lambda_3 & \mu_3 \end{vmatrix}}.$$

Our first task is to show that the cross ratio is well defined.

Theorem 1. *$R(p_1p_2p_3p_4)$ is independent of the parametrization.*

PROOF. Let p_i have parameters (λ_i, μ_i) relative to one parametrization, and parameters (λ_i', μ_i') relative to another parametrizatiom. By Exercise 1.3.10, there exists a non-singular 2×2 matrix M such that

$$(\lambda_i', \mu_i') = (\lambda_i, \mu_i)M \tag{1}$$

for each i. We shall employ the matrix identities

$$\begin{vmatrix} a & b \\ c & d \end{vmatrix} = (a,b)\begin{pmatrix} 0 & 1 \\ -1 & 0 \end{pmatrix}(c,d)^{\mathrm{T}} \tag{2}$$

and

$$\begin{pmatrix} a & b \\ c & d \end{pmatrix}\begin{pmatrix} 0 & 1 \\ -1 & 0 \end{pmatrix}\begin{pmatrix} a & b \\ c & d \end{pmatrix}^{\mathrm{T}} = \begin{vmatrix} a & b \\ c & d \end{vmatrix}\begin{pmatrix} 0 & 1 \\ -1 & 0 \end{pmatrix} \tag{3}$$

which you can easily verify by straightforward calculation. Combining (1), (2), and (3) appropriately, we have

$$\begin{aligned}
\begin{vmatrix} \lambda_i' & \mu_i' \\ \lambda_j' & \mu_j' \end{vmatrix} &= (\lambda_i', \mu_i')\begin{pmatrix} 0 & 1 \\ -1 & 0 \end{pmatrix}(\lambda_j', \mu_j')^{\mathrm{T}} \\
&= (\lambda_i, \mu_i)M\begin{pmatrix} 0 & 1 \\ -1 & 0 \end{pmatrix}[(\lambda_j, \mu_j)M]^{\mathrm{T}} \\
&= (\lambda_i, \mu_i)M\begin{pmatrix} 0 & 1 \\ -1 & 0 \end{pmatrix}M^{\mathrm{T}}(\lambda_j, \mu_j)^{\mathrm{T}} \\
&= |M|(\lambda_i, \mu_i)\begin{pmatrix} 0 & 1 \\ -1 & 0 \end{pmatrix}(\lambda_j, \mu_j)^{\mathrm{T}} \\
&= |M|\begin{vmatrix} \lambda_i & \mu_i \\ \lambda_j & \mu_j \end{vmatrix}.
\end{aligned}$$

Hence

$$\mathbb{R}(p_1p_2p_3p_4)=\frac{\begin{vmatrix}\lambda_1' & \mu_1'\\ \lambda_3' & \mu_3'\end{vmatrix}\begin{vmatrix}\lambda_2' & \mu_2'\\ \lambda_4' & \mu_4'\end{vmatrix}}{\begin{vmatrix}\lambda_1' & \mu_1'\\ \lambda_4' & \mu_4'\end{vmatrix}\begin{vmatrix}\lambda_2' & \mu_2'\\ \lambda_3' & \mu_3'\end{vmatrix}}$$

$$=\frac{|M|\begin{vmatrix}\lambda_1 & \mu_1\\ \lambda_3 & \mu_3\end{vmatrix}|M|\begin{vmatrix}\lambda_2 & \mu_2\\ \lambda_4 & \mu_4\end{vmatrix}}{|M|\begin{vmatrix}\lambda_1 & \mu_1\\ \lambda_4 & \mu_4\end{vmatrix}|M|\begin{vmatrix}\lambda_2 & \mu_2\\ \lambda_3 & \mu_3\end{vmatrix}}$$

$$=\frac{\begin{vmatrix}\lambda_1 & \mu_1\\ \lambda_3 & \mu_3\end{vmatrix}\begin{vmatrix}\lambda_2 & \mu_2\\ \lambda_4 & \mu_4\end{vmatrix}}{\begin{vmatrix}\lambda_1 & \mu_1\\ \lambda_4 & \mu_4\end{vmatrix}\begin{vmatrix}\lambda_2 & \mu_2\\ \lambda_3 & \mu_3\end{vmatrix}}.$$

Thus, cross ratio is the same in both parametrizations. □

It happens that if the points p_1, p_2, p_3, p_4 are taken in different orders, different cross ratios will result.

Theorem 2. *If $\mathbb{R}(p_1p_2p_3p_4)=r$, then $\mathbb{R}(p_1p_2p_4p_3)=1/r$ and $\mathbb{R}(p_1p_3p_2p_4)=1-r$.*

The proof is left as an exercise; straightforward calculation is all that is needed.

Just as we defined the cross ratio of four collinear points, we can dually define the cross ratio of four concurrent lines. The expression for the cross ratio is the same, but in terms of the parameters of the lines in a parametrization of the pencil. The following theorem then relates the two cross ratios.

Theorem 3. *If L_1, L_2, L_3, L_4 are four concurrent lines, T is a transversal of these lines, and $p_i = T \cap L_i$, then $\mathbb{R}(p_1p_2p_3p_4)=\mathbb{R}(L_1L_2L_3L_4)$.*

PROOF. Let L_3 and L_4 be base lines in the parametrization of the pencil, so that parameters of the lines are

$$L_1:\quad (\alpha_1, \beta_1),$$

$$L_2:\quad (\alpha_2, \beta_2),$$

$$L_3:\quad (1,0),$$

$$L_4:\quad (0,1).$$

Similarly, let p_3 and p_4 be base points on T, so that the parameters of the points are

$$\begin{aligned} p_1&: \ (\lambda_1, \mu_1),\\ p_2&: \ (\lambda_2, \mu_2),\\ p_3&: \ (1,0),\\ p_4&: \ (0,1). \end{aligned}$$

Also, let coordinates for the points be

$$p_i: \ [p_{i1}, p_{i2}, p_{i3}].$$

and let coordinates for the lines be

$$L_i: \ \langle l_{i1}, l_{i2}, l_{i3} \rangle.$$

Since p_i is on L_i, we have

$$\sum_{j=1}^{3} p_{ij} l_{ij} = 0$$

for each i. Now

$$p_1 = [\lambda_1 p_{31} + \mu_1 p_{41}, \lambda_1 p_{32} + \mu_1 p_{42}, \lambda_1 p_{33} + \mu_1 p_{43}]$$

and

$$L_1 = \langle \alpha_1 l_{31} + \beta_1 l_{41}, \alpha_1 l_{32} + \beta_1 l_{42}, \alpha, l_{33} + \beta_1 l_{43} \rangle;$$

since p_1 is on L_1, we have

$$\sum (\alpha_1 l_{3j} + \beta_1 l_{4j})(\lambda_1 p_{3j} + \mu_1 p_{4j}) = 0,$$

or

$$\sum (\alpha_1 \lambda_1 l_{3j} p_{4j} + \alpha_1 \mu_1 l_{3j} p_{4j} + \beta_1 \lambda_1 l_{4j} p_{3j} + \beta_1 \mu_1 l_{4j} p_{4j}) = 0.$$

Now

$$\sum l_{3j} p_{3j} = 0, \qquad \sum l_{4j} p_{4j} = 0,$$

so

$$\alpha_1 \mu_1 \sum l_{3j} p_{4j} + \beta_1 \lambda_1 \sum l_{4j} p_{3j} = 0.$$

Thus

$$(\alpha_1, \beta_1) = (\lambda_1 \sum l_{4j} p_{3j}, -\mu_1 \sum l_{3j} p_{4j}).$$

Similarly,

$$(\alpha_2, \beta_2) = (\lambda_2 \sum l_{4j} p_{3j}, -\mu_2 \sum l_{3j} p_{4j}).$$

Hence

$$\begin{aligned} \mathcal{R}(L_1 L_2 L_3 L_4) &= \frac{-\beta_1 \alpha_2}{-\alpha_1 \beta_2} \\ &= \frac{\mu_1 (\sum l_{3j} p_{4j}) \lambda_2 (\sum l_{4j} p_{3j})}{\lambda_1 (\sum l_{4j} p_{3j}) \mu_2 (\sum l_{3j} p_{4j})} \\ &= \frac{\mu_1 \lambda_2}{\lambda_1 \mu_2} \\ &= \mathcal{R}(p_1 p_2 p_3 p_4). \end{aligned}$$

□

It is Theorem 3 that enables us to prove that the cross ratio is an invariant, specifically an invariant under projectivities. For a perspectivity "preserves" the cross ratio of four points, by Theorem 3; that is, the cross ratio of the images of the four points is the same as the cross ratio of the four points. Since a projectivity is a composition of perspectivities, the same is true for a projectivity. We state this result as a theorem.

Theorem 4. *If there is a projectivity* $(abcd) \sim (a'b'c'd')$, *then* $R(a'b'c'd') = R(abcd)$.

The cross ratio can be used effectively in the Euclidean plane, where it has a particularly nice representation in terms of line segments. We may state the basic formula as a theorem.

Theorem 5. *In* α_R, *let* (xy) *denote the directed distance from point* x *to point* y. *If* p_1, p_2, p_3, p_4 *are four collinear points in* α_R, *then*

$$R(p_1p_2p_3p_4) = \frac{(p_1p_3)\cdot(p_2p_4)}{(p_1p_4)\cdot(p_2p_3)}.$$

PROOF. We may assume without loss of generality that a coordinate system in α_R is chosen so that p_1 is the origin and p_2 is the unit point $(1,0)$. In that coordinate system, let p_3 and p_4 have coordinates $(r,0)$ and $(s,0)$ respectively. Now viewing α_R as a subplane of π_R, the points have coordinates

$$\begin{aligned} p_1&: [0,0,1], \\ p_2&: [1,0,1], \\ p_3&: [r,0,1], \\ p_4&: [s,0,1]. \end{aligned}$$

Selecting the points $[1,0,0]$ and $[0,0,1]$ as base points, we find the parameters of these points to be

$$\begin{aligned} p_1&: (0,1), \\ p_2&: (1,1), \\ p_3&: (r,1), \\ p_4&: (s,1). \end{aligned}$$

Hence

$$R(p_1p_2p_3p_4) = \frac{-r(1-s)}{(1-r)(-s)}.$$

Now

$$(p_1p_3) = r, \quad (p_2p_4) = s-1, \quad (p_1p_4) = s, \quad (p_2p_3) = r-1.$$

Hence

$$R(p_1p_2p_3p_4) = \frac{(p_1p_3)\cdot(p_2p_4)}{(p_1p_4)\cdot(p_2p_3)},$$

as asserted. □

Harmonic sets are also easily characterized using the cross ratio. The basic theorem is as follows:

Theorem 6. *(ab, cd) is a harmonic set if and only if $\mathcal{R}(abcd) = -1$.*

PROOF. Suppose (ab, cd) is a harmonic set. Then there is a projectivity $(abcd) \sim (abdc)$ by Theorem 3.4.9. Hence by Theorem 4, $\mathcal{R}(abcd) = \mathcal{R}(abdc)$. But by Theorem 2, $\mathcal{R}(abdc) = \mathcal{R}(abcd)^{-1}$. Hence

$$\mathcal{R}(abcd)^2 = 1.$$

Now let a and b be chosen as base points, and let a, b, c, d have parameters

$$\begin{aligned} a&: \quad (1,0), \\ b&: \quad (0,1), \\ c&: \quad (\lambda, \mu), \\ d&: \quad (\lambda', \mu'). \end{aligned}$$

Then

$$\mathcal{R}(abcd) = \frac{\mu\lambda'}{\mu'\lambda}.$$

Now if $\mathcal{R}(abcd) = -1$, we are done. If $\mathcal{R}(abcd) = 1$, then

$$\mu\lambda' = \mu'\lambda,$$

so that

$$(\lambda', \mu') = (\lambda, \mu)$$

and

$$c = d.$$

Hence by Exercise 3.1.9, our plane is a Fano plane, so F has characteristic 2 and $1 = -1$ by Theorem 3.4.13. Hence $\mathcal{R}(abcd) = -1$ anyway. The proof of the converse is left as an exercise. □

Combining Theorems 5 and 6, we get an interpretation of a harmonic set in the Euclidean plane. (See, further, Court [6, Chapter 7].)

Theorem 7. *In α_R, (ab, cd) is a harmonic set if and only if c and d divide the segment $\overline{ab}$ internally and externally in the same ratio.*

PROOF. (ab, cd) is a harmonic set if and only if $(ac)(bd)/(ad)(bc) = -1$, if and only if $(ac)/(cb) = -(ad)/(db)$. This last expression means that c and d divide $\overline{ab}$ in the same ratio, and the negative sign indicates that one of c and d is between a and b and the other is not. □

Exercises 3.5

1. Compute $\mathcal{R}(p_1 p_2 p_3 p_4)$, given

   (a) parameters p_1: $(1,3)$, p_2; $(1,4)$, p_3: $(-1,1)$, p_4 $(2,3)$.
   (b) coordinates p_1: $[1,2,0]$, p_2: $[0,1,1]$, p_3: $[1,4,2]$, p_4: $[-2,-3,1]$.

***2.** Examine possible values of the cross ratio $\mathcal{R}(p_1p_2p_3p_4)$ in which p_1, p_2, p_3, p_4 are not distinct.

3. Prove Theorem 2.

***4.** State and prove the converse of Theorem 4.

5. Prove that $\mathcal{R}(abcd) = \mathcal{R}(badc) = \mathcal{R}(cdab) = \mathcal{R}(dcba)$.

6. Given $\mathcal{R}(abcd) = r$, find the cross ratios of each of the 24 permutations of a, b, c, d.

7. Complete the proof of Theorem 6.

8. Let $p_1 = [x_1, x_2, x_3]$, $p_2 = [y_1, y_2, y_3]$, $p_3 = [z_1, z_2, z_3]$, and $p_4 = [w_1, w_2, w_3]$. Prove that each of the expressions

$$\frac{\begin{vmatrix} x_1 & x_2 \\ z_1 & z_2 \end{vmatrix} \cdot \begin{vmatrix} y_1 & y_2 \\ w_1 & w_2 \end{vmatrix}}{\begin{vmatrix} x_1 & x_2 \\ w_1 & w_2 \end{vmatrix} \cdot \begin{vmatrix} y_1 & y_2 \\ z_1 & z_2 \end{vmatrix}}, \quad \frac{\begin{vmatrix} x_1 & x_3 \\ z_1 & z_3 \end{vmatrix} \cdot \begin{vmatrix} y_1 & y_3 \\ w_1 & w_3 \end{vmatrix}}{\begin{vmatrix} x_1 & x_3 \\ w_1 & w_3 \end{vmatrix} \cdot \begin{vmatrix} y_1 & y_3 \\ z_1 & z_3 \end{vmatrix}}, \quad \frac{\begin{vmatrix} x_2 & x_3 \\ z_2 & z_3 \end{vmatrix} \cdot \begin{vmatrix} y_2 & y_3 \\ w_2 & w_3 \end{vmatrix}}{\begin{vmatrix} x_2 & x_3 \\ w_2 & w_3 \end{vmatrix} \cdot \begin{vmatrix} y_2 & y_3 \\ z_2 & z_3 \end{vmatrix}}$$

that is defined is equal to the cross ratio $\mathcal{R}(p_1p_2p_3p_4)$. (Hence, we can use coordinates instead of parameters to compute cross ratios.)

***9.** If a, b, p, q, r are five collinear points, prove that

$$\mathcal{R}(abpq)\,\mathcal{R}(abqr)\,\mathcal{R}(abrp) = 1.$$

***10.** Give a meaningful answer to these questions:

(a) What is the harmonic conjugate of p relative to p and q?
(b) What is the harmonic conjugate of p relative to q and q?

11. In α_R, show that the harmonic conjugate of the ideal point on line ab relative to a and b is the midpoint of segment $\overline{ab}$.

12. Let L_1 and L_2 be two intersecting lines in α_R. Let L_3 and L_4 be the two angle bisectors of the angles formed by L_1 and L_2. Let T be any transversal of L_1, L_2, L_3, and L_4, and let $p_i = T \cap L_i$. Prove that (p_1p_2, p_3p_4) is a harmonic set.

13. Let γ_1 and γ_2 be two circles in α_R that do not intersect, and let c_1 and c_2 be their centers. Let a common external tangent of γ_1 and γ_2 meet c_1c_2 at c_3, and let a common internal tangent meet c_1c_2 at c_4. Prove that (c_1c_2, c_3c_4) is a harmonic set. (Hint: Use internal and external division.)

14. Use Exercise 13 to devise a construction for the common internal and external tangents of two nonintersecting circles in α_R.

15. In π_c, let $j_1 = [1, i, 0]$ and $j_2 = [1, -i, 0]$. Given lines L and M in α_R, let l and

m be the ideal points on L and M. Prove that $L \perp M$ (in the usual sense) if and only if $(j_1 j_2, lm)$ is a harmonic set.

16. Prove that for four collinear points a, b, c, d in α_R, (ab, cd) is a harmonic set if and only if $2/(ab) = 1/(ac) + 1/(ad)$, if and only if (ac), (ab), (ad) form a harmonic progression.

17. Prove that a collineation of π_R is projective if and only if it preserves cross ratio.

18. Let f be a harmonic homology on π_R with center c and axis A. For any point p, let $a = pc \cap A$. Prove that $(ca, pf(p))$ is a harmonic set. Hence devise a construction for the image of any point under f, given only c and A.

Chapter 4

Conics in Pappian Planes

In this chapter, we study the conic sections, or *conics*, from the projective viewpoint. We have already referred to conics in the Euclidean plane α_R and their extensions in π_R and π_C (Sections 1.7, 2.6); here we develop conics in a more general setting.

Throughout this chapter, we consider only Pappian planes. Thus, any projective plane mentioned can be regarded as the plane π_F for some field F.

Section 4.1. The Projective Definition of a Conic

In Sections 2.1 and 2.2, we developed properties of perspectivities and projectivities between ranges. Further properties of projectivities in Pappian planes were developed in Section 3.4.

By dualizing these results, we obtain the properties of perspectivities and projectivities between pencils. We can also talk about projectivities between ranges and pencils, as follows.

Let L be a range and let p be a pencil. The *perspectivity* between L and p is the function $f: L \barwedge p$ such that for each x on L, $f(x) = px$. A *projectivity* between L and p is a function $f: L \sim p$ such that there is a range L' with $f: L \sim L' \barwedge p$.

Dually, we can define a projectivity between a pencil p and a range L; it is just the inverse of $L \sim p$.

If p and q are pencils, we shall denote the projectivity $p(pa, pb, pc, \dots)$ $\sim q(qa, qb, qc, \dots)$ between them by $p(a,b,c,\dots) \sim q(a,b,c,\dots)$. Here it should be clear that $a,b,c,\dots$ need not be collinear. Dually, we shall denote the projectivity $L(L \cap A, L \cap B, \dots) \sim M(M \cap A, M \cap B, \dots)$ between ranges L and M by $L(A,B,\dots) \sim M(A,B,\dots)$.

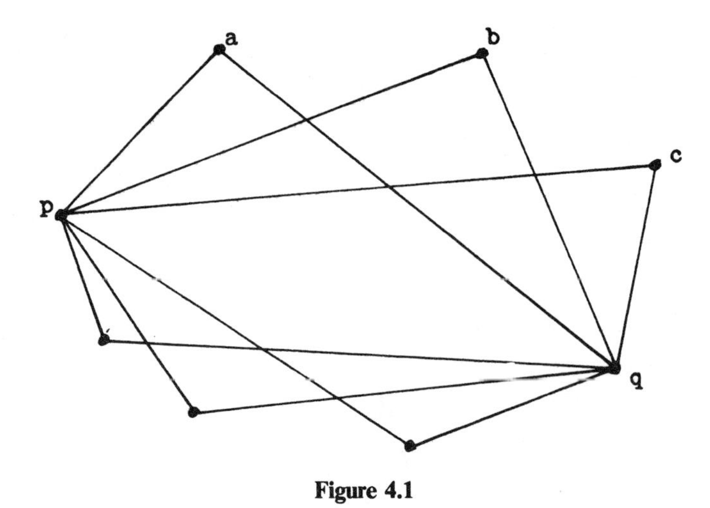

Figure 4.1

Now we are ready to give a projective definition of a conic. This definition is due to J. Steiner (1796–1862); see Coolidge [5, p. 97].

Definition 1. Let $f: p \sim q$ be a projectivity between pencils p and q (f not the identity if $p = q$). The *point conic* Γ_f *generated* by f is the locus of points x such that $f(px) = qx$. Points p and q are called *generating bases* of Γ_f. If $p = q$ or if f is a perspectivity, we call Γ_f *singular*.

Thus a point conic is the locus of intersections of corresponding lines under a projectivity between pencils (see Figure 4.1). Thus if $a, b, c, \ldots$ are

Figure 4.2

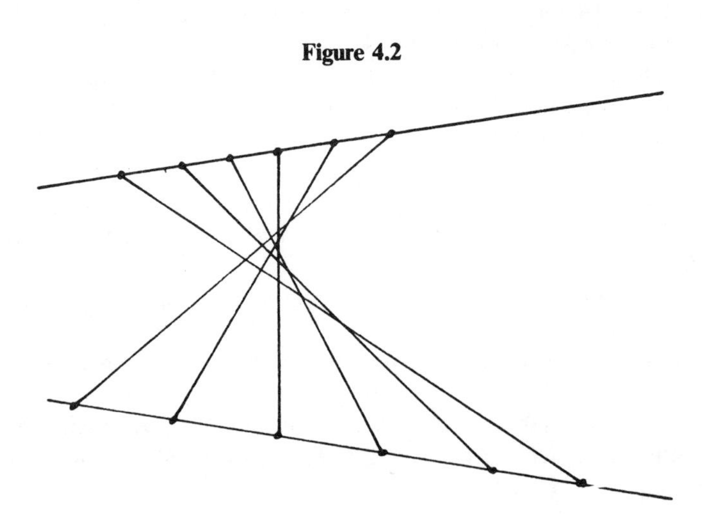

points of the point conic with generating bases p and q, then $p(a,b,c,\ldots)$ $\sim q(a,b,c,\ldots)$.

The dual of a point conic is a *line conic*, and is the envelope of lines joining corresponding points under a projectivity between ranges (see Figure 4.2). In what follows, we shall refer primarily to point conics; similar results for line conics are found using duality.

If the point conic Γ_f is singular, there are several possible cases. If $p = q$, then f can have zero, one, or two fixed lines (by the dual of the fundamental theorem 3.4.5). If f has no fixed lines, then Γ_f consists of the single point p $(= q)$; if f has one fixed line, then Γ_f consists of a single range, the range on the fixed line; if f has two fixed lines, Γ_f consists of these two ranges. If $p \neq q$ and f is a perspectivity, then line pq is self-corresponding (by the dual of the definition of perspectivity) and the corresponding lines under f meet on a line, called the *axis* of the perspectivity. In this case, Γ_f consists of two ranges, pq and the axis.

If the point conic Γ_f is not singular (or *nonsingular*), then f is not a perspectivity and the generating bases p and q are distinct. It is not hard to prove that a point conic contains its generating bases.

It is also easy to prove the following theorem; the proof is left as an exercise.

Theorem 1. *If p,q,a,b,c are five points, no three of which are collinear, then there is a unique point conic with generating bases p and q and containing a, b, and c; the conic is nonsingular.*

Not quite so easily proved is the following theorem.

Theorem 2. *Any two points of a nonsingular point conic may be taken as generating bases.*

PROOF. Suppose Γ_f is a nonsingular point conic with generating bases p and q. Then $p \neq q$. Let $a,b,c,d,e,\ldots$ be any other points on Γ_f. We shall show that a and b can be chosen as generating bases by proving that there exists a projectivity $g: a(pqcde\cdots)\sim b(pqcde\cdots)$ so that $\Gamma_f = \Gamma_g$. To do so, let $L = cd$, $a' = pa \cap L$, $b' = pb \cap L$, $a'' = qa \cap L$, $b'' = qb \cap L$ (see Figure 4.3). Then we have

$$\begin{aligned} L(a'b'cd) &\barwedge p(abcd) \\ &\sim q(abcd) && \text{by definition of } \Gamma_f \\ &\barwedge L(a''b''cd) \\ &\sim L(b''a''dc) && \text{by Theorem 3.4.8.} \end{aligned}$$

Thus $(a'b'cd)\sim(b''a''dc)$ is a projectivity on L that interchanges c and d. Therefore by Exercise 3.4.10, this projectivity is an involution; hence it interchanges a' and b'', b' and a'', by Exercise 2.2.7. That is, $(a'a''b'b''cd)\sim(b''b'a''a'dc)$, or in particular,

$$\begin{aligned} (a'a''cd) &\sim (b''b'dc) \\ &\sim (b'b''cd) \end{aligned}$$

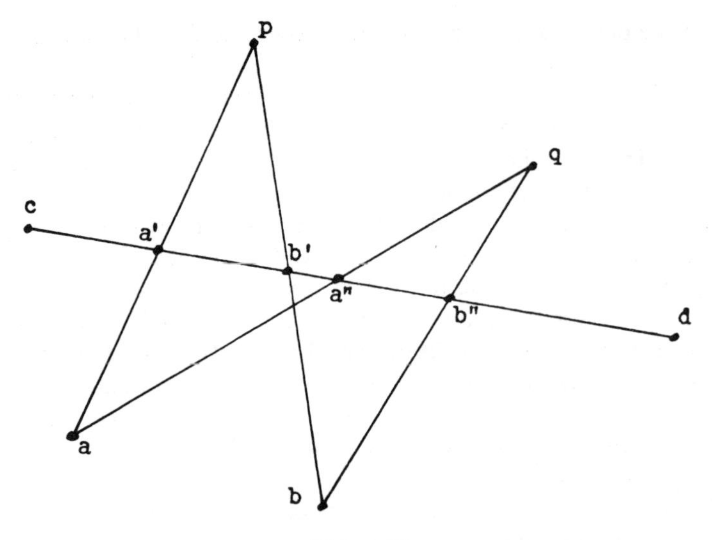

Figure 4.3

by the permutation theorem again. Hence

$$\begin{aligned} a(pqcd) &\sim L(a'a''cd) \\ &\sim L(b'b''cd) \\ &\sim b(pqcd). \end{aligned}$$

In a similar fashion, $a(pqce) \sim b(pqce)$, so that $a(pqcde) \sim b(pqcde)$. Let $g: a(pqcde \cdots) \sim b(pqcde \cdots)$. Then $\Gamma_g = \Gamma_f$, as claimed. □

Exercises 4.1

1. Give the definition of a perspectivity between pencils.

*2. State the dual of the fundamental theorem, Theorem 3.4.5.

3. State the dual of the perspectivity theorem, Theorem 3.4.4.

4. Give the possible kinds of singular line conics.

5. Prove that a point conic contains its generating bases.

6. Prove Theorem 1.

7. Let p, q, a, b, c be five points, no four of which are collinear. Under what conditions is there a unique point conic with generating bases p and q that contains a, b, and c? Is the conic singular or nonsingular?

8. Is there always a point conic containing five given points?

*9. Given five points, no three of which are collinear, prove that there is a unique point conic containing the five points.

Section 4.2. Intersections of a Range and a Point Conic

We now bring up the question of how a range may be related to a point conic. We have already seen that a singular point conic may contain a range.

Our first theorem states that a nonsingular conic cannot contain a range.

Theorem 1. *A nonsingular point conic may contain at most two points of a given range.*

PROOF. Suppose range L intersects a point conic Γ in three points a, b, and c. If Γ contains two points p and q not on L, then $f: p(abc) \sim q(abc)$ is a perspectivity, and $\Gamma = \Gamma_f$ is singular. If Γ contains only one point p not on L, then using p and a as generating bases, we have $p(bc \ldots) \sim a(bc \ldots)$, and pb and pc both correspond to L, an impossibility. If Γ consists only of points on L, then using a and b as generating bases, we have $a(c \cdots) \sim b(c \cdots)$; but then every line through a must correspond to line $ba = L$, an impossibility. Thus if Γ is nonsingular, L meets Γ in at most two points. □

A line on two points of a nonsingular point conic is called a *secant line* of the conic. A line on exactly one point of a nonsingular point conic is called a *tangent line* of the conic, and the point of intersection is called the *point of contact* or *point of tangency*. A line on no point of a nonsingular point conic is called a *nonintersector* of the conic.

Dually, a point on two lines of a nonsingular line conic is an *exterior point*, and a point on exactly one line is a *tangent point*. A point on no lines of a nonsingular line conic is called an *interior point* of the conic.

The relationship of tangent lines to the generation of a point conic is given by the next theorem.

Theorem 2. *If Γ_f is a nonsingular point conic with generating bases p and q, then $f(pq)$ is the line tangent to Γ_f at q, and $f^{-1}(pq)$ is the line tangent to Γ_f at p (see Figure* 4.4).

PROOF. Line pq meets its image line $f(pq)$ at q; thus $f(pq)$ meets Γ_f at q, but at no other point, since $f(pq)$ is the image of pq alone. Thus $f(pq)$ is tangent to Γ_f. A similar argument applies to $f^{-1}(pq)$. □

The existence of nonintersectors of a point conic and of interior points of a line conic depends on properties of the projective plane.

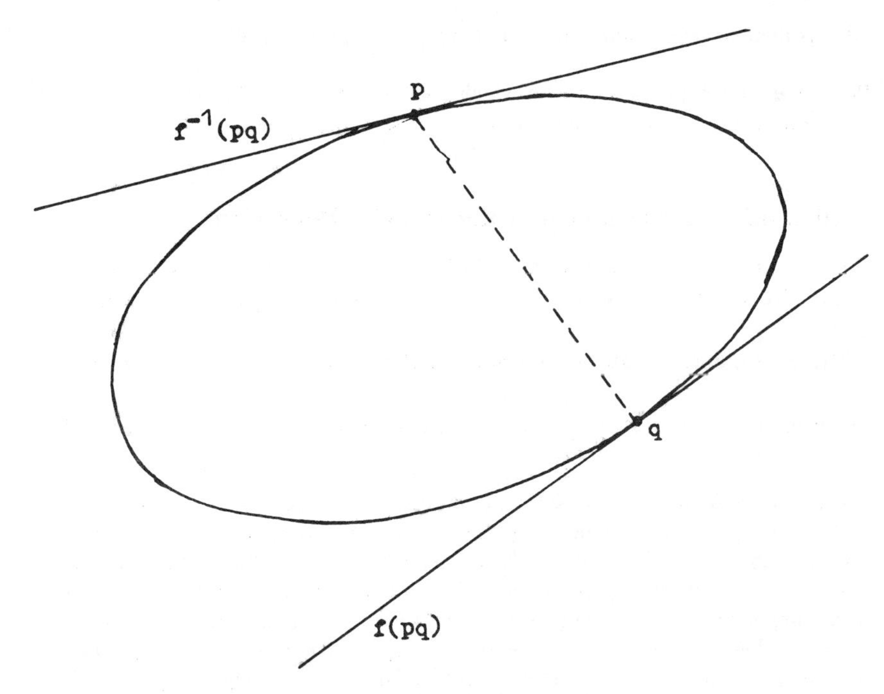

Figure 4.4

Definition 1. A projective plane is called *closed* if and only if every projectivity $f: L \sim L$ from a range to itself has a fixed point.

The first consequence of closure of a projective plane is the following.

Theorem 3. *In a closed Pappian plane, every line meets every nonsingular point conic.*

PROOF. Let Γ_f be a point conic with generating bases p and q. Let L be any line. Define $g: L \barwedge p \overset{f}{\sim} q \barwedge L$; then g is a projectivity from range L to itself, so it has a fixed point s. But then ps corresponds to qs under f, so s belongs to Γ_f. Thus L meets Γ_f. □

Thus in a closed Pappian plane, a nonsingular point conic has no nonintersectors. As you will see in the exercises, it is also true that a singular point conic has no nonintersectors; moreover, no line conic has interior points in a closed plane.

Experience with the Euclidean plane, in which a circle has nonintersectors, for example, shows us that the real projective plane is not closed. This is because in the real number system, not every number has a square root, so not every quadratic equation has a solution. A field in which every quadratic equation has a solution is called *quadratically closed*. The general case can now be stated as follows.

Theorem 4. *The plane π_F is closed if and only if the field F is quadratically closed.*

PROOF. Let projectivity $f: L \sim L$ in π_F have equation

$$f(\lambda, \mu) = (\lambda', \mu') = (\lambda, \mu)\begin{pmatrix} a & b \\ c & d \end{pmatrix}.$$

Then f has a fixed point if and only if

$$(\lambda, \mu)\begin{pmatrix} a & b \\ c & d \end{pmatrix} = k(\lambda, \mu),$$

if and only if

$$(\lambda, \mu)\begin{pmatrix} a-k & b \\ c & d-k \end{pmatrix} = (0,0),$$

if and only if

$$\begin{vmatrix} a-k & b \\ c & d-k \end{vmatrix} = 0,$$

if and only if

$$k^2 - (a+d)k + (ad - bc) = 0,$$

which has a solution if and only if F is quadratically closed. □

Thus, for example, π_C is a closed plane.

Exercises 4.2

*1. State the dual of Theorem 1.

*2. Prove that if p is a point on a nonsingular point conic Γ, then exactly one line through p is tangent to Γ.

*3. State the dual of the result of Problem 2.

4. Given five points, no three of which are collinear, construct the line on one of them that is tangent to the point conic they determine.

*5. State the dual of Theorem 2.

6. Carry out the dual of Problem 4.

*7. Prove that in a closed projective plane, every projectivity $f: p \sim p$ from a pencil to itself has a fixed line.

*8. Prove that the principle of duality holds in the class of closed Pappian planes.

9. Show that in a closed Pappian plane, a singular point conic consists of one or two ranges.

*10. Prove that in a closed Pappian plane, every line meets every point conic.

11. Show that every element of the field Z_2 has a square root, but Z_2 is not quadratically closed.

12. Show that in a closed plane, an involution always has two fixed points.

Section 4.3. Conics in a Closed Plane π_F

In this section, we shall assume that the plane π_F is closed, or equivalently, that F is quadratically closed. Our aim is to discuss conics analytically, or in terms of the algebra in F.

Our first theorem gives us the equation of a point conic.

Theorem 1. *A point conic in π_F is the locus of points $[x_1, x_2, x_3]$ satisfying an equation of the form*

$$ax_1^2 + bx_2^2 + cx_3^2 + 2dx_1x_2 + 2ex_1x_3 + 2fx_2x_3 = 0.$$

PROOF. Let Γ_f be a point conic in π_F with generating bases p and p'. Let $L = \langle l_1, l_2, l_3 \rangle$ and $M = \langle m_1, m_2, m_3 \rangle$ be base lines on p, and let $L' = \langle l_1', l_2', l_3' \rangle$ and $M' = \langle m_1', m_2', m_3' \rangle$ be base lines on p'. If the line with parameters (λ, μ) on p, namely $\langle \lambda l_1 + \mu m_1, \lambda l_2 + \mu m_2, \lambda l_3 + \mu m_3 \rangle$, has as its image under f the line $\langle \lambda' l_1' + \mu' m_1', \lambda' l_2' + \mu' m_2', \lambda' l_3' + \mu' m_3' \rangle$ with parameters (λ', μ'), then f has an equation of the form

$$\alpha\lambda\lambda' + \beta\lambda\mu' + \gamma\mu\lambda' + \delta\mu\mu' = 0 \qquad (*)$$

for some $\alpha, \beta, \gamma, \delta \in F$.

Now suppose point $x = [x_1, x_2, x_3]$ belongs to Γ_f. Then x is on some line of pencil p, and also on its image line in p'. Hence both

$$\sum_{i=1}^{3} (\lambda l_i + \mu m_i)x_i = 0$$

and

$$\sum_{i=1}^{3} (\lambda' l_i' + \mu' m_i')x_i = 0.$$

These equations may be rewritten as

$$\lambda(\textstyle\sum l_i x_i) + \mu(\sum m_i x_i) = 0,$$
$$\lambda'(\textstyle\sum l_i' x_i) + \mu'(\sum m_i' x_i) = 0.$$

Solving, we may select

$$\lambda = \textstyle\sum m_i x_i, \qquad \mu = -\sum l_i x_i$$

and

$$\lambda' = \textstyle\sum m_i' x_i, \qquad \mu' = -\sum l_i' x_i.$$

Substituting these expressions into (*), we get

$$\alpha(\textstyle\sum m_i x_i)(\sum m_i' x_i) + \beta(\sum m_i x_i)(-\sum l_i' x_i)$$
$$+\gamma(-\textstyle\sum l_i x_i)(\sum m_i' x_i) + \delta(-\sum l_i x_i)(-\sum l_i' x_i) = 0,$$

which can be expanded and rewritten in the form given in the theorem. Thus if point x belongs to the conic, then it satisfies the equation. By reversing the above steps, we can see that any point satisfying the equation must belong to the conic. □

Dually, we have the equation of a line conic.

Theorem 2. *A line conic in π_F is the envelope of lines $\langle l_1, l_2, l_3\rangle$ satisfying an equation of the form*

$$al_1^2 + bl_2^2 + cl_3^2 + 2dl_1l_2 + 2el_1l_3 + 2fl_2l_3 = 0.$$

These equations can be written in a more compact matrix form, which has advantages in many settings. If we let x represent not only a point but also the row vector (x_1, x_2, x_3) of coordinates of the point, then the equation

$$ax_1^2 + bx_2^2 + cx_3^2 + 2dx_1x_2 + 2ex_1x_3 + 2fx_2x_3 = 0$$

can be written

$$xAx^{\mathrm{T}} = 0,$$

with x^{T} the transpose of x and

$$A = \begin{bmatrix} a & d & e \\ d & b & f \\ e & f & c \end{bmatrix}.$$

This is verified by straightforward computation.

Dually, if L is the row vector of coordinates of a line, then the equation of a line conic is of the form

$$LAL^{\mathrm{T}} = 0.$$

Example 1. In π_R, the point conic

$$x_1^2 + 4x_2^2 - x_3^2 + 6x_1x_2 - 4x_1x_3 - x_2x_3 = 0$$

has equation $xAx^{\mathrm{T}} = 0$ with

$$A = \begin{bmatrix} 1 & 3 & -2 \\ 3 & 4 & -\frac{1}{2} \\ -2 & -\frac{1}{2} & -1 \end{bmatrix}.$$

If $\Gamma : xAx^{\mathrm{T}} = 0$ is a point conic in π_F and p is any point, the line L with equation

$$pAx^{\mathrm{T}} = 0$$

is called the *polar* of p relative to Γ. If $pA = (0,0,0)$, the polar of p is indeterminate, and we call p a *vertex* of Γ.

***Example* 2.** For the conic $\Gamma: xAx^{\mathrm{T}} = 0$ of Example 1, the polar of $p = [2, -1, 3]$ is the line with equation

$$(2,-1,3)\begin{bmatrix} 1 & 3 & -2 \\ 3 & 4 & -\frac{1}{2} \\ -2 & -\frac{1}{2} & -1 \end{bmatrix}\begin{bmatrix} x_1 \\ x_2 \\ x_3 \end{bmatrix} = 0,$$

or

$$\left(-8, -\tfrac{7}{2}, -\tfrac{13}{2}\right)\begin{bmatrix} x_1 \\ x_2 \\ x_3 \end{bmatrix} = 0,$$

or

$$-8x_1 - \tfrac{7}{2}x_2 - \tfrac{13}{2}x_3 = 0.$$

The coordinates of this line are

$$\left\langle -8, -\tfrac{7}{2}, -\tfrac{13}{2}\right\rangle,$$

or, equally well,

$$\langle 16, 7, 13\rangle.$$

We see from Example 2 that if p is not a vertex of $\Gamma: xAx^{\mathrm{T}} = 0$, then the polar of p has coordinate vector pA.

The idea of the polar of a point with respect to a conic is useful in describing the conic. For example, a singular point conic is one with a vertex, as the next sequence of theorems shows.

Theorem 3. *The point conic* $\Gamma: xAx^{\mathrm{T}} = 0$ *has a vertex if and only if the matrix* A *is singular.*

PROOF. If v is a point, the equation

$$vA = (0,0,0)$$

can be regarded as a homogeneous system of three linear equations in three variables. Hence Γ has a vertex v if and only if v is a nontrivial solution of $vA = (0,0,0)$, which happens if and only if the determinant $|A|$ is zero, if and only if A is singular. □

Theorem 4. *If the point conic* $\Gamma: xAx^{\mathrm{T}} = 0$ *is singular, then* Γ *has a vertex.*

PROOF. If Γ is singular, then Γ consists of one or two ranges, by Exercise 4.2.9. Let $L = \langle l_1, l_2, l_3\rangle$ and $M = \langle m_1, m_2, m_3\rangle$ be the ranges, where possibly $L = M$. Then the equation of Γ is

$$(l_1x_1 + l_2x_2 + l_3x_3)(m_1x_1 + m_2x_2 + m_3x_3) = 0,$$

or $xAx^{\mathrm{T}} = 0$ with

$$A = \begin{bmatrix} l_1m_1 & \frac{1}{2}(l_1m_2 + l_2m_1) & \frac{1}{2}(l_1m_3 + l_3m_1) \\ \frac{1}{2}(l_1m_2 + l_2m_1) & l_2m_2 & \frac{1}{2}(l_2m_3 + l_3m_2) \\ \frac{1}{2}(l_1m_3 + l_3m_1) & \frac{1}{2}(l_2m_3 + l_3m_2) & l_3m_3 \end{bmatrix}.$$

Now we may verify directly (by tedious calculation) that $|A| = 0$, so Γ has a vertex by Theorem 3. Alternatively, if $L \neq M$ and

$$v = \left[\begin{vmatrix} l_2 & l_3 \\ l_2 & l_3 \end{vmatrix}, \begin{vmatrix} l_3 & l_1 \\ m_3 & m_1 \end{vmatrix}, \begin{vmatrix} l_1 & l_2 \\ m_1 & m_2 \end{vmatrix} \right],$$

then we may verify that $vA = (0,0,0)$ (again, the calculations are tedious). If $L = M$, then

$$|A| = \begin{vmatrix} l_1^2 & l_1l_2 & l_1l_3 \\ l_1l_2 & l_2^2 & l_2l_3 \\ l_1l_3 & l_2l_3 & l_3^2 \end{vmatrix} = l_1l_2l_3 \begin{vmatrix} l_1 & l_2 & l_3 \\ l_1 & l_2 & l_3 \\ l_1 & l_2 & l_3 \end{vmatrix}$$

and Theorem 3 applies again. □

Theorem 5. *If $\Gamma : xAx^{\mathrm{T}} = 0$ is a point conic, v is a vertex of Γ, and p is a point of Γ, then Γ contains the range of pv.*

PROOF. Let $x = \lambda p + \mu v$ be any point of the range pv. (We are using p to represent both a point and its row vector of coordinates.) Then

$$\begin{aligned} xAx^{\mathrm{T}} &= (\lambda p + \mu v)A(\lambda p + \mu v)^{\mathrm{T}} \\ &= \lambda^2 pAp^{\mathrm{T}} + \lambda\mu pAv^{\mathrm{T}} + \mu\lambda vAp^{\mathrm{T}} + \mu^2 vAv^{\mathrm{T}}. \end{aligned}$$

Now $pAp^{\mathrm{T}} = 0$, because p is on Γ. Also vAp^{T} and vAv^{T} are zero, because $vA = (0,0,0)$. The matrix pAv^{T} is a 1×1 matrix, so $pAv^{\mathrm{T}} = (pAv^{\mathrm{T}})^{\mathrm{T}} = vA^{\mathrm{T}}p^{\mathrm{T}} = vAp^{\mathrm{T}}$, since A is symmetric, and $vAp^{\mathrm{T}} = 0$. Hence $xAx^{\mathrm{T}} = 0$, and Γ contains x. Thus Γ contains pv. □

Corollary. *If a point conic Γ has a vertex, then Γ is singular.*

Thus, combining these results, we see that a point conic $\Gamma : xAx^{\mathrm{T}} = 0$ is singular if and only if A is singular, if and only if Γ has a vertex.

To describe further properties of conics and polars, we need the following lemma.

Lemma 1. *Let line pq meet the point conic $\Gamma : xAx^{\mathrm{T}} = 0$ in π_F at point x. Then the parameters (λ, μ) of x relative to the base points p and q satisfy the equation*

$$(pAp^{\mathrm{T}})\lambda^2 + 2(pAq^{\mathrm{T}})\lambda\mu + (qAq^{\mathrm{T}})\mu^2 = 0.$$

PROOF. Let $x = \lambda p + \mu q$. Since x is on Γ, we have

$$(\lambda p + \mu q)A(\lambda p + \mu q)^T = 0,$$
$$\lambda^2 pAp^T + \lambda\mu pAq^T + \mu\lambda qAp^T + \mu^2 qAq^T = 0,$$

which reduces to the required equation because A is symmetric and

$$qAp^T = (qAp^T)^T = pA^Tq^T = pAq^T. \qquad \square$$

Now we can investigate polars relative to nonsingular conics.

Theorem 6. *If $\Gamma : xAx^T = 0$ is a nonsingular point conic and p is a point on Γ, then the polar of p relative to Γ is the line on p that is tangent to Γ.*

PROOF. The polar of p is the line $pAx^T = 0$; since p is on Γ, we have $pAp^T = 0$, and p lies on its own polar. Let q be any point on the polar of p, distinct from p. Then $pAq^T = 0$. Now we apply Lemma 1: Any point on both Γ and the polar pq has parameters (λ, μ) satisfying

$$(pAp^T)\lambda^2 + 2(pAq^T)\lambda\mu + (qAq^T)\mu^2 = 0, \qquad (**)$$

which reduces to

$$(qAq^T)\mu^2 = 0.$$

If $qAq^T = 0$, then Equation (**) is satisfied by every pair (λ, μ), and Γ contains range pq. But that is impossible, by Theorem 4.2.1. Hence $qAq^T \neq 0$, and therefore $\mu = 0$. Thus the intersection of pq with Γ is the point $\lambda p + 0q = p$, and the polar of p is tangent to Γ at p. $\square$

Theorem 7. *If $\Gamma : xAx^T = 0$ is a nonsingular point conic and p is a point not on Γ, then the polar of p relative to Γ is the line on the two points of contact of the tangents from p to Γ.*

PROOF. Since every line meets every conic, let q be a point on Γ and on the polar of p. Then $qAq^T = 0$ and $pAq^T = 0$. Hence by Lemma 1 any point on pq and Γ has parameters (λ, μ) satisfying

$$(pAp^T)\lambda^2 = 0.$$

Since p is not on Γ, $pAp^T \neq 0$, so $\lambda = 0$. Thus only q is on both pq and Γ, so pq is tangent to Γ. Note that the polar line is not on p, and so is not tangent to Γ; hence the polar of p meets Γ in two points, and there are two lines on p tangent to Γ (see Figure 4.5). $\square$

Further properties of polars will be pursued in the exercises and in the next section. We now pause to dualize the above results.

If $\Gamma : LAL^T = 0$ is a line conic, and M is any line, the point with equation

$$MAL^T = 0,$$

or coordinate vector MA, is called the *pole* of M relative to Γ. If $MA = (0, 0, 0)$, the line M is called a *side* of Γ.

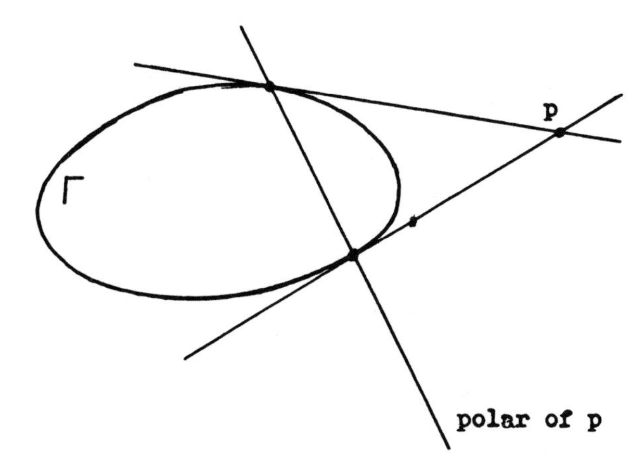

Figure 4.5

The duals of Theorems 3–5 tell us that a line conic is singular if and only if it has a side, if and only if its matrix is singular. The dual of Theorem 6 says that the pole of a line belonging to a nonsingular line conic is the tangent point on that line. The dual of Theorem 7 says that the pole of a line not belonging to a nonsingular line conic is the point on the two lines of the conic that pass through the two tangent points of the conic on the original line.

If you understand that last sentence, you are doing well. Intuitively, point conics are easier to deal with than line conics; tangent lines seem to make more sense than tangent points. For these reasons, we seek to relate point and line conics, so that we can think of them in the same terms. The basic theorem is the following.

Theorem 8. *If $\Gamma : xAx^{\mathrm{T}} = 0$ is a nonsingular point conic, then the envelope of lines tangent to Γ is the line conic $\Gamma' : LA^{-1}L^{\mathrm{T}} = 0$.*

PROOF. We first show that any line L tangent to Γ satisfies the equation $LA^{-1}L^{\mathrm{T}} = 0$. If L is tangent to Γ at point p, then L is the polar of p, or $L = pA$. Hence

$$\begin{aligned} LA^{-1}L^{\mathrm{T}} &= (pA)A^{-1}(pA)^{\mathrm{T}} \\ &= pAA^{-1}A^{\mathrm{T}}p^{\mathrm{T}} \\ &= pA^{\mathrm{T}}p^{\mathrm{T}} \\ &= pAp^{\mathrm{T}} = 0, \end{aligned}$$

as claimed. Now we show that each line of $\Gamma' : LA^{-1}L^{\mathrm{T}} = 0$ is tangent to Γ. If L is an arbitrary line of Γ', let $LA^{-1} = p$. Then $pAp^{\mathrm{T}} = (LA^{-1})A(LA^{-1})^{\mathrm{T}} = LA^{-1}AA^{-1\mathrm{T}}L^{\mathrm{T}} = LA^{-1}L^{\mathrm{T}} = 0$, so p is a point of Γ. Finally, $pA = (LA^{-1})A = L$, so L is the polar of p, and thus is tangent to Γ. □

We call the line conic Γ' in Theorem 8 the *derived conic* of Γ. The value of the derived conic is that it enables us to talk about poles in terms of a point conic.

Theorem 9. *If $\Gamma : xAx^{\mathrm{T}} = 0$ is a nonsingular point conic, p is a point, and L is the polar of p relative to Γ, then p is the pole of L relative to the derived conic $\Gamma' : LA^{-1}L^{\mathrm{T}} = 0$.*

PROOF. The pole of L is $LA^{-1} = (pA)A^{-1} = p$. □

Because of this relationship, we call p the pole of L *relative to* Γ, rather than relative to Γ'. Thus a point and line are "pole and polar" relative to Γ when the line is the polar of the point or the point is the pole of the line.

We can also combine the point conic Γ and its derived conic into a single self-dual figure; henceforth, when we say conic, we shall mean both a point conic and its envelope of tangent lines. In this setting, the dual of Theorem 7 makes more sense.

Exercises 4.3

1. Show that if a point Γ in π_C contains the two points $[1, i, 0]$ and $[1, -i, 0]$, then the Euclidean specialization of Γ is a circle.

2. State and prove the converse of Exercise 1.

3. Write the equation
$$2x_1^2 + 3x_2^2 + x_3^2 - 2x_1x_2 + 2x_1x_3 = 0$$
in matrix form.

4. Find the polars of $p = [1, 0, -1]$, $q = [2, 2, 3]$, and $r = [5, 5, -9]$ relative to the conic in Exercise 3.

5. For the conic $xAx^{\mathrm{T}} = 0$ with
$$A = \begin{pmatrix} 2 & 0 & -3 \\ 0 & -2 & -3 \\ -3 & -3 & 0 \end{pmatrix}$$
show that $v = [3, -3, 2]$ is a vertex.

6. Find a vertex of $\Gamma : xAx^{\mathrm{T}} = 0$ in π_c for
$$\text{(a)}\ A = \begin{pmatrix} 1 & 2 & -1 \\ 2 & 0 & 2 \\ -1 & 2 & -3 \end{pmatrix}, \quad \text{(b)}\ A = \begin{pmatrix} 0 & 4 & 3 \\ 4 & 0 & 0 \\ 3 & 0 & 0 \end{pmatrix}, \quad \text{(c)}\ A = \begin{pmatrix} 1 & 0 & 1 \\ 0 & 0 & 0 \\ 1 & 0 & 1 \end{pmatrix}.$$

*7. Prove that if Γ is a singular point conic with vertex v, then the polar of any point p relative to Γ passes through v.

8. Prove that if u and v are both vertices of Γ, then every point of range uv is a vertex of Γ.

9. Show that the conic $\Gamma : xAx^{\mathrm{T}} = 0$ is nonsingular if A has rank 3, consists of two ranges if A has rank 2, and consists of one range if A has rank 1.

10. In π_c, let the conic Γ have equation $x_1^2 - x_2x_3 = 0$. Find the lines tangent to Γ which pass through the points (a) $[0,0,1]$, (b) $[1,0,0]$, (c) $[1,1,-1]$, (d) $[0,1,0]$.

***11.** A triangle is said to be *self-polar* relative to a point conic Γ in case the polar of each vertex of the triangle relative to Γ is the opposite side. Show that $\triangle pqr$ in Exercise 4 is self-polar relative to the conic of Exercise 3.

12. Find the most general point conic relative to which the triangle with vertices $[1,0,0]$, $[0,1,0]$, and $[0,0,1]$ is self-polar.

13. Discuss the derived conic of a singular point conic.

14. For the conic in Exercise 3, find the pole of line $\langle 1,2,-1\rangle$.

15. Dualize the definition of self-polar triangle found in Exercise 11.

16. If $\triangle pqr$ is self-polar relative to Γ, is it also self-polar relative to the derived conic Γ'?

17. Let point p have polar L relative to Γ, and let line M on p meet L at q and Γ at r and s. Show that (pq, rs) is a harmonic set.

18. Show that if q is on the polar of p relative to Γ, then p is on the polar of q.

19. In case the plane π_F is not closed, a conic Γ may have nonintersectors and interior points. Use Exercise 18 to construct

(a) the polar of an interior point,
(b) the pole of a nonintersector.
(Hint: Use a circle in α_R to get some ideas.)

Section 4.4. Desargues's Conic Theorem

In this section, we shall work in a closed Pappian plane satisfying Fano's axiom.

The main theorem of this section is due to Desargues, and is called his conic theorem to distinguish it from his triangle theorem. Before stating it, we present a related result.

Theorem 1. *If a_0, a_1, a_2, a_3 are four points on a nonsingular conic Γ, then the diagonal triangle of the four-point $a_0a_1a_2a_3$ is self-polar relative to Γ. (See Figure 4.6.)*

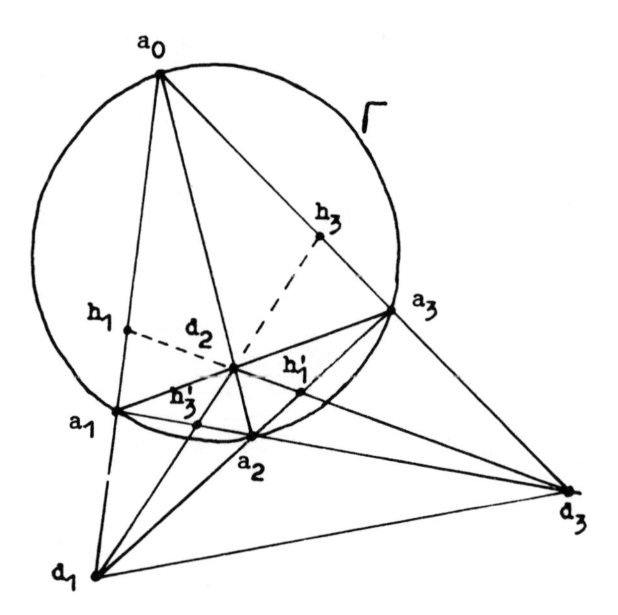

Figure 4.6

PROOF. The polar of d_1 passes through the harmonic conjugates of d_1 relative to a_0 and a_1 and to a_2 and a_3, by Exercise 4.3.17. Now (d_1d_2, h_3h_3') is a harmonic set, by Exercise 3.1.10, and $(d_1d_2h_3h_3') \overset{d_3}{\barwedge} (d_1h_1a_0a_1) \sim (a_0a_1d_1h_1)$ by the permutation theorem. Hence h_1 is the harmonic conjugate of d_1 relative to a_0 and a_1. Similarly, h_1' is the harmonic conjugate of d_1 relative to a_2 and a_3. Thus $h_1h_1' = d_2d_3$ is the polar of d_1. Similar results hold for d_2 and d_3, so that the diagonal triangle $d_1d_2d_3$ is self-polar. □

One immediate consequence of Theorem 1 is that there exist straightedge constructions of polars and tangent lines in α_R; see the exercises.

Now to Desargues's conic theorem.

Theorem 2. Desargues's Conic Theorem. *If $a_0a_1a_2a_3$ is a four-point in a closed Pappian plane satisfying Fano's axiom and L is a line not on any a_i, then there is an involution $f\colon L \sim L$ such that any point conic on a_0, a_1, a_2, and a_3 meets L in a pair of mates under f.*

PROOF. Let $x_i = L \cap a_0a_i$, $i = 1, 2, 3$, and let $x_1' = L \cap a_2a_3$, $x_2' = L \cap a_1a_3$, $x_3' = L \cap a_1a_2$ (see Figure 4.7). Let f be the projectivity $f\colon L(x_1x_2x_3) \sim L(x_1'x_2'x_3')$. Note that

$$L(x_1x_2x_3x_1') \overset{a_0}{\barwedge} (d_1a_2a_3x_1') \sim (x_1'a_3a_2d_1) \overset{a_1}{\barwedge} L(x_1'x_2'x_3'x_1),$$

so that f interchanges x_1 and x_1'. Thus by Exercise 3.4.10, f is an involution.

Now suppose a conic Γ on a_0, a_1, a_2, a_3 meets L at points y and z. Then let g be

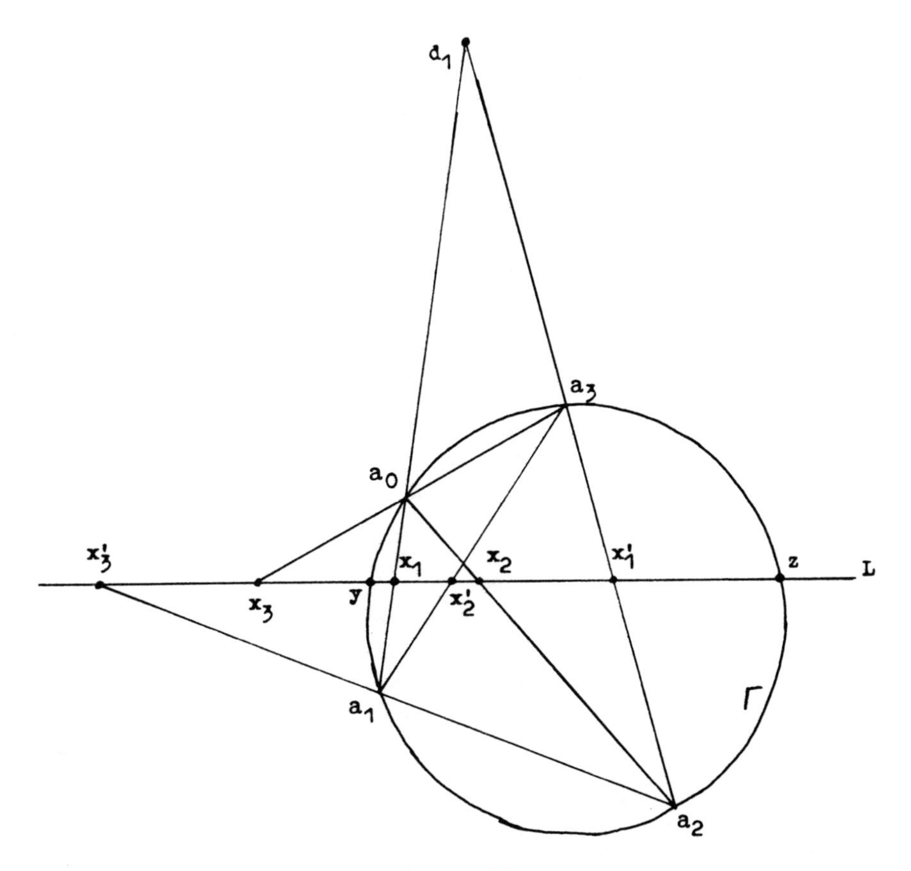

Figure 4.7

the projectivity

$$g : L(x_1x_2yz) \barwedge a_0(a_1a_2yz) \sim a_3(a_1a_2yz)$$
$$\barwedge L(x_3'x_1'yz) \sim L(x_1'x_2'zy),$$

and let h be the projectivity

$$h : L(x_1x_3yz) \barwedge a_0(a_1a_3yz) \sim a_x(a_1a_3yz)$$
$$\barwedge L(x_3'x_1'yz) \sim L(x_1'x_3'zy).$$

Now both g and h carry x_1 to x_1', y to z, and z to y; hence $g = h$ and $L(x_1x_2x_3yz) \sim L(x_1'x_2'x_3'zy)$. Hence $g = h = f$, and Γ meets L in a pair of mates under f. □

As an application of Desargues's conic theorem, we can say that there are always exactly two conics which pass through a given four-point and are tangent to a given line. That is because any conic through the four points meets the line in points that are mates in an involution; but if the conic is tangent to the line, the single point of intersection must be a fixed point of the involution. Since the involution must have exactly two fixed points, exactly two conics satisfy the requirements.

Another application of Desargues's conic theorem is that it gives a point-by-point construction of a conic. That is, if five points on the conic are known, let L be a line through one of them. Then the other four points induce an involution on L, and the image of the fifth point under that involution is the second intersection of L with the conic. By varying L, as many points of the conic can be constructed as desired.

We shall state for reference two special cases of Desargues's conic theorem, leaving the proofs as exercises.

Theorem 3. *If pqr is a triangle and T is a line on r but not on p or q, and L is a line not on p, q, or r, then there is an involution $f: L \sim L$ such that any point conic through p and q and tangent to T at r meets L in a pair of mates under f.*

Theorem 4. *If p and q are two points, S is a line on p but not q, and T is a line on q but not p, and L is any line not on p or q, then there is an involution $f: L \sim L$ such that any point conic that is tangent to S and p and tangent to T at q meets L in a pair of mates under f.*

Exercises 4.4

1. Given a circle in α_R and a point not on the circle, devise a straightedge construction for the polar of the point relative to the circle. [Hint: Use Theorem 1.]

2. Given a circle in α_R and an exterior point, devise a straightedge construction for the tangents to the circle from the point. (N.B.: A straightedge can only be used to draw the line joining two points.)

3. Given four points in α_R, how many parabolas pass through the four points? Do there exist four points through which no parabola passes? [Hint: Consider closure and ideal points.]

4. Using Desargues's conic theorem, devise a point-by-point construction of a conic, given five points, no three collinear.

5. Prove Theorem 3.

6. Prove Theorem 4.

Section 4.5. Pascal's Theorem

In 1641, Blaise Pascal proved (at age 16) one of the most beautiful theorems in geometry. We call it Pascal's theorem, but its Euclidean version is sometimes called the theorem of the mystic hexagram. In this section, we present Pascal's theorem and its applications.

We continue to work in a closed Pappian plane satisfying Fano's axiom.

Theorem 1. Pascal's Theorem. *If a, b, c, a', b', and c' are six points on a nonsingular conic Γ, then the points*

$$a'' = bc' \cap b'c, \qquad b'' = ac' \cap a'c, \qquad c'' = ab' \cap a'b$$

are collinear.

PROOF. Let $L = ab'$ and $M = ac'$. Also let $b_1 = L \cap bc'$ and $c_1 = M \cap cb'$ (see Figure 4.8). Then

$$\begin{aligned} L(ac''b'b_1) &\barwedge b(aa'b'c') \\ &\sim c(aa'b'c') \\ &\barwedge M(ab''c_1c'). \end{aligned}$$

Hence $L(ac''b'b_1) \sim M(ab''c_1c')$ is a perspectivity with center $b'c_1 \cap b_1c' = a''$. Therefore $b''c''$ is on a''. □

In the Euclidean setting, a *hexagram* is a generalized hexagon, whose sides may intersect at points other than vertices. If the six points a, b, c, a',

Figure 4.8

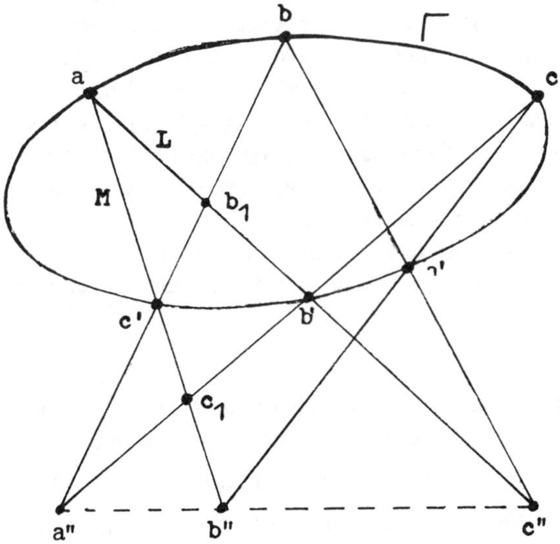

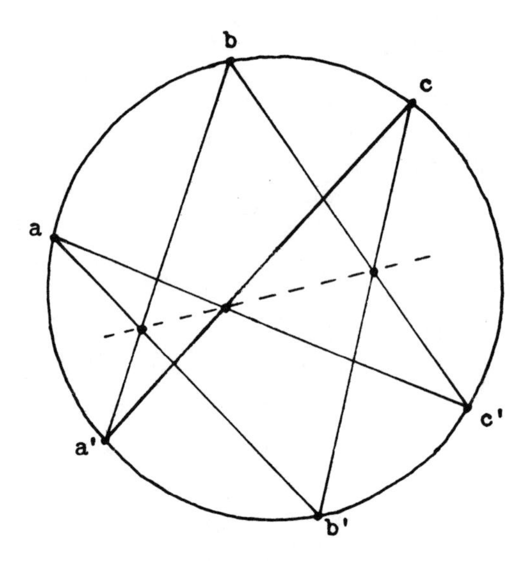

Figure 4.9

b', c' are joined in the order $ab'ca'bc'a$, we obtain a hexagram, and the pairs of sides yielding points a'', b'', and c'' are pairs of opposite sides. Thus the "theorem of the mystic hexagram" is that if a hexagram is inscribed in a circle, then the pairs of opposite sides meet in three collinear points. (See Figure 4.9).

The line containing a'', b'', and c'' is called the *Pascal line* for the hexagram $ab'ca'bc'$. It should be evident that for different orderings of the six points we get different hexagrams and hence different Pascal lines.

The dual of Pascal's theorem is known as Brianchon's theorem, for C. J. Brianchon (1785–1864), who proved it in 1806. Interestingly enough, the principle of duality was not formulated until some twenty years later, by J.-V. Poncelet (1788–1867), J.-D. Gergonne (1771–1859), and J. Plucker (1801–1868). Pascal's and Brianchon's theorems provided one of the first examples of nontrivial dual theorems.

You are invited to state Brianchon's theorem.

While Pascal's theorem is indeed beautiful, the converse of Pascal's theorem is more useful .

Theorem 2. Converse of Pascal's Theorem. *If a, b, c, a', b', and c' are six points, no three of which are collinear, and the points*

$$a'' = bc' \cap b'c, \qquad b'' = ac' \cap a'c, \qquad c'' = ab' \cap a'b$$

are collinear, then the six points all lie on a nonsingular conic.

PROOF. Let Γ be the conic on points a, b, c, a', and b'. Let line ac' meet Γ for the

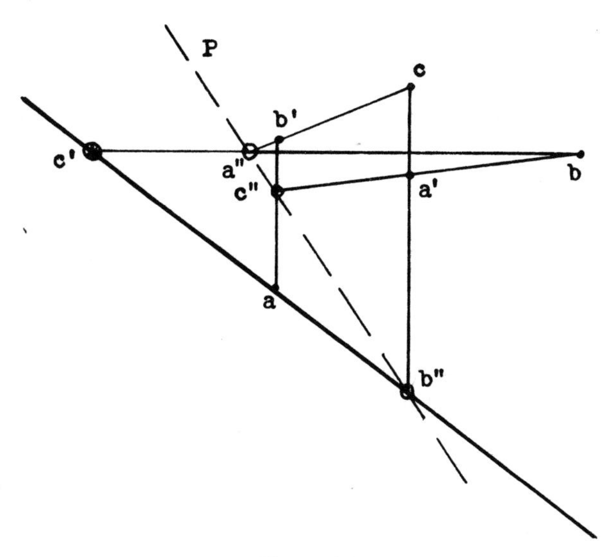

Figure 4.10

second time at c^*. Then points

$$a^* = bc^* \cap b'c, \qquad b^* = ac^* \cap a'c, \qquad c'' = ab' \cap a'b$$

are collinear by Pascal's theorem. Now $ac^* = ac'$, so $b^* = b''$. Thus a^*, b'', and c'' are collinear, so

$$a^* = b''c'' \cap b'c.$$

But by hypothesis,

$$a'' = b''c'' \cap b'c,$$

so $a'' = a^*$. Therefore bc^* passes through a'', but also bc' passes through a''. Thus c^* is on $ba'' = bc'$, so $c^* = bc' \cap ac' = c'$, and the theorem is proved. □

The utility of the converse of Pascal's theorem is that it provides a simple point-by-point construction of a conic. Given five points, no three collinear, and a line on one of them, we find the second intersection of the line with the conic on the five points as follows.

Let a be the point on the line (see Figure 4.10), and let b', c, a', and b be the other four points. Let c' be the second point on the line and the conic; we seek to construct c'. We first construct $c'' = ab' \cap a'b$ and $b'' = ac' \cap a'c$. (Even though we do not know c', we do know line ac'.) This gives us the Pascal line $P = b''c''$ of $ab'ca'bc'$. Then we construct $a'' = b'c \cap P$; then $c' = ba'' \cap ac'$.

Now by varying the line through a, we can construct as many points on the conic as we like.

Pascal's theorem also has some special cases that should be mentioned. We shall state these cases without proof, and then discuss their significance.

Theorem 3. *If a, b, c, a', and b' are five points on a nonsingular conic Γ and T is the line tangent to Γ at a, then the points*

$$a'' = ba \cap b'c, \qquad b'' = T \cap a'c, \qquad c'' = ab' \cap a'b$$

are collinear.

Theorem 4. *If a, b, c, and a' are four points on a nonsingular conic Γ, S is the line tangent to Γ at a, and T is the line tangent to Γ at b, then the points*

$$a'' = T \cap ac, \qquad b'' = ab \cap a'c, \qquad c'' = S \cap a'b$$

are collinear.

The easy way to remember the results of Theorems 3 and 4 is that points of tangency are *double labeled* with adjacent points in the ordering $ab'ca'bc'$, and then Pascal's theorem is applied, replacing the line that names the same point twice with the tangent line. Thus in Theorem 3, we simply set $c' = a$, and replace the reference to line ac' with line T. We also replace bc' with ba. In Theorem 4, we have $a = b'$ and $b = c'$, and replace ab' with S and bc' with T.

The converse of Theorems 3 and 4 also hold and provide for constructions of conics determined by four points and the tangent at one of them, or by three points and the tangents at two of them. The same double-labeling idea works in the constructions.

Exercises 4.5

1. How many different Pascal lines are associated with six points on a conic? How are they related?

2. Explain how the theorem of Pappus might be viewed as a special case of Pascal's theorem.

3. State Brianchon's theorem in terms of a hexagram circumscribed about a circle.

4. State the converse of Theorem 3.

5. State the converse of Theorem 4.

6. Given four points, no three collinear, and a line on one of them, let L be another line on one of the points. Devise a construction for the second intersection of L with the conic on the four points and tangent to the given line.

7. Given five points, no three collinear, a unique conic is determined. Devise a construction for the line tangent to the conic at one of the given points.

8. Given three points of a nonsingular conic and the tangent lines at two of them, devise a construction for the second intersection of a line through one of the given points with the conic.

9. Given four points of a nonsingular conic and the tangent line at one of them, construct the tangent line at another of the points.

10. State the special case of Pascal's theorem for the case of three points on a nonsingular conic and the tangent lines at all three of them.

11. Given three points of a nonsingular conic and the tangent lines at two of them, construct the tangent line at the third point.

12. In α_R, let triangle abc be inscribed in a circle Γ, and let lines A, B, C be tangent to Γ at a, b, c respectively. If $a' = B \cap C$, $b' = A \cap C$, and $c' = A \cap B$, show that triangles abc and $a'b'c'$ form a central couple.

13. In α_R, given three points of a parabola and the direction of its axis, construct the vertex, the axis, the focus, and the directrix.

14. In α_R, given the asymptotes of a hyperbola and one point p on the hyperbola, construct the line tangent to the hyperbola at p.

15. From Exercise 14, show that the segment of a tangent to a hyperbola terminated by the asymptotes is bisected by the point of tangency.

16. Given five points, no three collinear, and a sixth point p, devise a construction of the polar of p relative to the conic on the five points.

Section 4.6. Polarities

Our definition of a conic in terms of a projectivity between pencils is due to Steiner. The theory of poles and polars relative to a conic is then a major part of the development.

But conics can be approached in the reverse direction. C. G. C. von Staudt (1798–1867) developed a theory of poles and polars, from which he obtained a conic as a certain configuration in that theory. In this section, we give a very brief outline of the theory of polarities, and show how they give rise to conics.

We start with the notion of a correlation.

Definition 1. Let $\pi = (\mathcal{P}, \mathcal{L}, \mathcal{I})$ be a Pappian plane. A *correlation* on π is a pair of bijections $f: \mathcal{P} \to \mathcal{L}$ and $g: \mathcal{L} \to \mathcal{P}$ such that if point p is on line L, then line $f(p)$ is on point $g(L)$.

Thus a correlation carries points to lines and lines to points in such a way that incidence is preserved dually. That is, a correlation is an isomorphism between π and its dual. Therefore a correlation carries ranges to pencils and pencils to ranges. The inverse of a correlation is also a correlation, and the product of correlations is a collineation.

A *projective* correlation is one in which any range is projective with its image pencil. It can be proved that if some range is projective with its image pencil, then the correlation is projective. It can also be shown that there is a projective correlation carrying a given four-point onto a given four-line. In these and other respects, the theory of correlations parallels the theory of collineations.

We can now define a polarity as a special kind of correlation.

Definition 2. A *polarity* is a projective correlation of period two.

That is, if a polarity carries point x to line X, then it also carries line X to point x. Thus the composition of a polarity with itself is the identity collineation.

An interesting result is the following.

Theorem 1. *Let abc be a triangle with sides* $A = bc$, $B = ca$, $C = ab$. *Any projective correlation that carries a to A, b to B, and c to C is a polarity.*

PROOF. Let p be a point not on A, B, or C, and let P be a line not on a, b, or c. Then there is a projective correlation carrying the four-point $abcp$ to the four-line $ABCP$. It will be sufficient to show that the correlation also carries $ABCP$ to $abcp$.

Figure 4.11

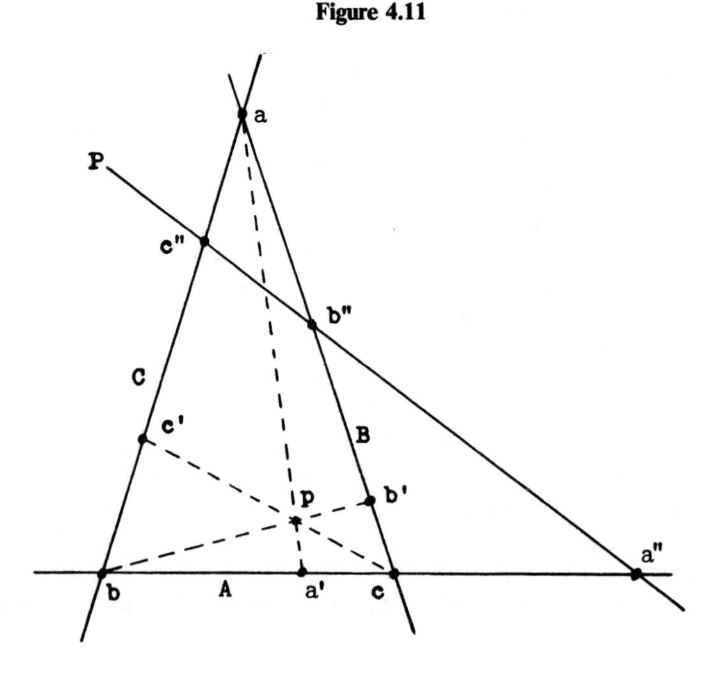

Since $A = bc$, the image of A is $B \cap C = a$, and similarly for the other two sides of the triangle. To show that P is carried to p, let $a' = ap \cap A$, $b' = bp \cap B$, $c' = cp \cap C$, $a'' = P \cap A$, $b'' = P \cap B$, and $c'' = P \cap C$ (see Figure 4.11).

To find the image of a'', consider the range A, whose image is the pencil a. Let any point x on A have image X on a, and let $X \cap A = y$. Then the mapping from A to itself that carries x to y is a projectivity. Under this projectivity, points b and c are interchanged, so the projectivity is an involution. Now the correlation carries ap to $A \cap P = a''$, so it carries $a' = ap \cap A$ to $a''a$; hence the involution on A carries a' to a'', and therefore carries a'' to a'. Similarly, involutions on B and C carry b'' to b' and c'' to c'. Thus the correlation carries a'' to $aa' = ap$, b'' to $bb' = bp$, and c'' to $cc' = cp$. Hence the correlation carries $P = a''b''c''$ to $ap \cap bp \cap cp = p$, as required. Thus the correlation is a polarity. □

Under a polarity the image X of a point x is called the *polar* of x, and x is called the *pole* of its polar. If the vertices of a triangle are the poles of the opposite sides, the triangle is called *self-polar*. Thus Theorem 1 says that a triangle abc and a point p and line P, not related to the triangle, determine a polarity under which abc is self-polar.

Many properties of polarities can be developed at this point, but only those related to conics will be mentioned here. See the exercises for a few other examples.

Under a polarity, suppose point p has polar P. If q is a point on P, then the polar Q of q must pass through p. That is, if q is on the polar of p, then p is on the polar of q. In this setting, we call p and q *conjugate points*, and we also call P and Q *conjugate lines*.

If it happens that point p lies on its own polar P, then p and P are both *self-conjugate*. We can make the following statements about self-conjugacy.

Theorem 2. *If p and q are self-conjugate points, then line pq is not self-conjugate.*

PROOF. If $pq = L$ were self-conjugate, then its pole l must lie on L. Since p is on L, P must pass through l. However, p is on P, so $P = pl = L$. Similarly, we can show $Q = L$. Thus p and q would have the same polar, contradicting the fact that a polarity is one-to-one. □

Theorem 3. *If line L is not self-conjugate under a given polarity, then the polarity induces on L an involution of conjugate points, in which fixed points are self-conjugate.*

PROOF. If L is not self-conjugate, its pole l is not on L. If x is a point on L, its polar X lies on l; let $L \cap X = x'$. Since x' is on the polar of x, it follows that x and x' are conjugate points. Thus the mapping

$$f : L(xx' \cdots) \sim l(XX' \cdots) \barwedge L(x'x \cdots)$$

is an involution of conjugate points. That fixed points in this involution are self-conjugate is obvious. □

Corollary 1. *A line can pass through at most two self-conjugate points.*

PROOF. If the line is self-conjugate, it has only one self-conjugate point, by Theorem 2. If the line is not self-conjugate, it has at most two self-conjugate points, for an involution has at most two fixed points by the fundamental theorem. □

Corollary 2. *In a closed Pappian plane, every line contains one or two self-conjugate points.*

PROOF. If a line is self-conjugate, it has one self-conjugate point. If a line is not self-conjugate, it contains two self-conjugate points, for in a closed plane an involution has exactly two fixed points. □

Corollary 3. *On a line with two self-conjugate points, any two conjugate points are harmonic conjugates with respect to the self-conjugate points.*

PROOF. Use Exercise 3.4.12. □

Corollary 4. *In a closed Pappian plane, every point lies on one or two self-conjugate lines.*

PROOF. Dual of Corollary 2. □

Corollary 5. *In a closed Pappian plane, the polar of a non-self-conjugate point p is the line joining the poles of the two self-conjugate lines on p.*

PROOF. If p is not self-conjugate, two self-conjugate lines L and M lie on p by Corollary 4. Since p is on L and M, its polar P is on the poles l and m. □

We now begin to see that the self-conjugate points under a polarity form a point conic, and the self-conjugate lines are the tangents to the conic. The lines containing two self-conjugate points are just secants of the conic. Pole and polar under the polarity are the same as pole and polar relative to the conic.

We shall show that this approach does indeed lead to conics by proving that the self-conjugate points under a polarity satisfy the projective definition of a conic.

Theorem 4. Steiner's Theorem. *Let p and q be two self-conjugate points under a polarity, and define the mapping $f: p \to q$ from the pencil p to the pencil q as follows: If x is a self-conjugate point different from q, set $f(px) = qx$. If $x = q$, set $f(px) = Q$, the polar of q. Finally, set $f(P) = qp$. Then f is a projectivity.*

PROOF. Let $c = P \cap Q$, and let L be a line on c but not on p or q. For x any self-conjugate point other than p or q, let $a = px \cap L$ and $b = qx \cap L$ (see Figure 4.12). Let $c' = pq \cap L$. Since pq is the polar of c (by Corollary 5) and c is on L, the pole l of L must be on pq. Moreover, l and c' are conjugate points, so l is the

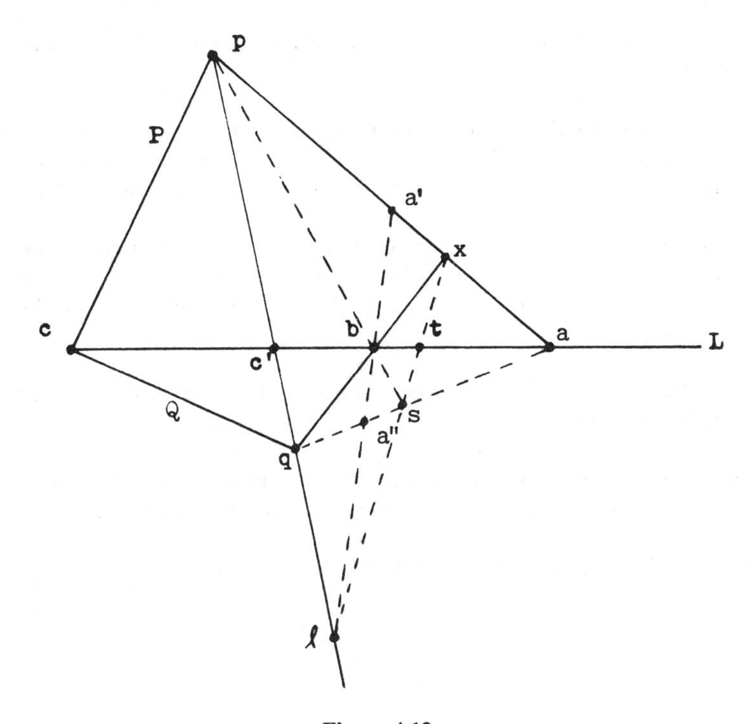

Figure 4.12

harmonic conjugate of c' relative to p and q, by Corollary 3. Using the four-point $abpq$ to construct l, let $s = pb \cap qa$; then $l = pq \cap xs$. Let $xs \cap L = t$; then (xs, lt) is a harmonic set. Since l and t are conjugate points, and are harmonic conjugates with respect to x and s, it follows by the uniqueness of harmonic conjugates that s is self-conjugate. Now let $a' = lb \cap px$ and $a'' = lb \cap qs$. Then (px, aa') and (qs, aa'') are harmonic sets. Thus by Corollary 3, lb is the polar of a, so that a and b are conjugate points. Hence, in the involution on L, we have

$$L(cc'ab) \sim L(c'cba).$$

Finally, we have

$$f : p(P, pq, px) \barwedge L(cc'a) \sim L(c'cb) \barwedge q(qp, Q, qx),$$

so that f is a projectivity. □

Exercises 4.6

1. Let M be a 3×3 matrix over F. Define f on π_F by $f(x) = \langle xM \rangle$, so that the image of point x is a line. Show that f defines a projective correlation.

2. If a triangle is not self-polar under a polarity, show that the triangle and its image ("polar") triangle form a central couple. [This is *Chasles's theorem*, after M. Chasles (1793–1880).]

3. In the setting of Problem 2, show that the center and axis of the couple are pole and polar.

4. If triangle abc is self-polar under a polarity, and point p not on a side of abc is not self-conjugate, show that the involution induced on P by the polarity is the same involution as induced by the four-point $abcp$ (see Exercise 3.4.11).

5. Let A and B be nonconjugate lines under a polarity. Show that every point x on A has a conjugate point y on B, and that the mapping $x \to y$ is a projectivity. Show that this projectivity is a perspectivity if and only if point $A \cap B$ is self-conjugate.

6. Prove that if the vertices of a four-point are self-conjugate, then the diagonal triangle is self-polar.

7. Prove *Seydewitz's theorem*: If points p, q, and r are self-conjugate and line L is conjugate to line pq, then $L \cap pr$ and $L \cap qr$ are conjugate points.

8. Under what conditions on M in Exercise 1 will f define a polarity?

Chapter 5

Metric Projective Geometry in π_C

Thus far in our study of projective geometry, the ideas of distance and angle have been conspicuously absent. Now we shall introduce metric ideas in a particular projective plane, namely π_C, the complex projective plane.

Once metric ideas are available, we may suitably specialize the metric to obtain analytic versions of the classical non-Euclidean geometries and even Euclidean geometry from π_C. This discovery in 1872 led A. Cayley (1821–1891) to exclaim, "All geometry is projective geometry!" Although we can't really agree with Cayley, his excitement was certainly justified. Modern developments in topology and some abstract geometries go beyond projective geometry, but most geometric ideas are included in projective geometry in one way or another.

Section 5.1. Distance and Angle in π_C

The usual Euclidean ideas of distance and angle do not extend to π_R, let alone to π_C. Therefore, to define a metric on π_C, we must start from scratch.

In general, a *metric* comes in two parts, a *distance relation* and an *angle relation*. From our experience in Euclidean geometry, we would expect a distance relation to have certain properties: A point should be at zero distance from itself; the distance from point p to point q should be the negative of the distance from q to p; and for collinear points p, q, and r, the distances between two pairs of the points should add up to the distance between the third pair. The angle relation has similar properties, complicated a bit by the fact that angles in α_R do not have unique measures.

Guided by these general principles, we are led to the following definition.

Definition 1. Let Γ be a nonsingular conic in π_C, viewed as a self-dual object containing a point conic and its envelope of tangent lines. A *metric* on π_C with *absolute* Γ is a *distance relation* $d_\Gamma : \mathcal{P} \times \mathcal{P} \to C$ and an *angle relation* $D_\Gamma : \mathcal{L} \times \mathcal{L} \to C$ defined as follows:
If p and q are not on Γ, let line pq meet Γ at points x and y, and set

$$d_\Gamma(p,q) = k_\Gamma \ln R(xypq).$$

If L and M are not in Γ, let point $L \cap M$ lie on lines X and Y of Γ, and set

$$D_\Gamma(L,M) = K_\Gamma \ln R(XYLM).$$

Points and lines of Γ are called *isotropic*. The constants k_Γ and K_Γ are called *scale constants*.

The basic properties of a metric on π_C are contained in the following theorem.

Theorem 1. *If (d_Γ, D_Γ) is a metric on π_C, if p, q, r, are nonisotropic points, and if L, M, N are nonisotropic lines, then*

1. $d_\Gamma(p,p) = 2nk_\Gamma \pi i$,
2. $d_\Gamma(q,p) = -d_\Gamma(p,q)$,
3. *If p, q, and r are collinear, then*
$$d_\Gamma(p,q) + d_\Gamma(q,r) + d_\Gamma(r,p) = (2n+1)k_\Gamma \pi i, \qquad n \text{ an integer},$$
4. $D_\Gamma(L,L) = 2nK_\Gamma \pi i$,
5. $D_\Gamma(M,L) = -D_\Gamma(L,M)$,
6. *If L, M, and N are concurrent, then*

$$D_\Gamma(L,M) + D_\Gamma(M,N) + D_\Gamma(N,L) = (2n+1)K_\Gamma \pi i, \qquad n \text{ an integer}.$$

PROOF. First, recall that for a complex number z, if $z = |z|e^{i\theta}$, then $\ln z = \ln|z| + i\theta$ (see Exercise 1).

To prove part 1, we use the fact that $R(xypp) = 1$ (Exercise 3.5.2), and get

$$\begin{aligned} d_\Gamma(p,p) &= k_\Gamma \ln R(xypp) \\ &= k_\Gamma \ln 1 \\ &= k_\Gamma[0 + i(2n\pi)] \\ &= 2nk_\Gamma \pi i. \end{aligned}$$

To prove part 2, we use the fact that $R(xyqp) = 1/R(xypq)$ (Theorem 3.5.2) and get

$$\begin{aligned} d_\Gamma(q,p) &= k_\Gamma \ln R(xyqp) \\ &= k_\Gamma \ln[R(xypq)]^{-1} \\ &= -k_\Gamma \ln R(xypq) \\ &= -d_\Gamma(p,q). \end{aligned}$$

To prove part 3, we use Exercise 3.5.9 and get

$$\begin{aligned} d_\Gamma(p,q) + d_\Gamma(q,r) + d_\Gamma(r,p) &= k_\Gamma \ln R(xypq) + k_\Gamma \ln R(xyqr) + k_\Gamma \ln R(xyrp) \\ &= k_\Gamma \ln[R(xypq)R(xyqr)R(xyrp)] \\ &= k_\Gamma \ln(-1) \\ &= k_\Gamma[0 + i(\pi + 2n\pi)] \\ &= (2n+1)k_\Gamma \pi i. \end{aligned}$$

Statements 4, 5, and 6 are the duals of 1, 2, and 3. □

We do not attempt to define distance between isotropic points, or between an isotropic point and a nonisotropic point. For nonisotropic points that lie on an isotropic line, we have the following surprising result:

Theorem 2. *If p and q are nonisotropic points on an isotropic line, then*

$$d_\Gamma(p,q) = 2nk_\Gamma \pi i.$$

PROOF. If p and q lie on an isotropic line, which is a tangent line, then only one point of intersection of line pq with Γ exists. Hence

$$\begin{aligned} d_\Gamma(p,q) &= k_\Gamma \ln R(xxpq) \\ &= k_\Gamma \ln 1 = 2nk_\Gamma \pi i. \end{aligned}$$

□

If the absolute Γ consists of the point conic $xAx^{\mathrm{T}} = 0$ and the line conic $LA^{-1}L^{\mathrm{T}} = 0$, then we may express distance and angle analytically in terms of coordinates. To simplify our calculations, if p and q are points (or row vectors of coordinates of points, to be precise) we let

$$f_{pq} = pAq^{\mathrm{T}}.$$

For lines L and M, we let

$$F_{LM} = LA^{-1}M^{\mathrm{T}}.$$

Theorem 3. *If p and q are nonisotropic points, collinear with the isotropic points x and y, then*

$$R(xypq) = \frac{-f_{pq} - \sqrt{f_{pq}^2 - f_{pp}f_{qq}}}{-f_{pq} + \sqrt{f_{pq}^2 - f_{pp}f_{qq}}}.$$

and $d_\Gamma(p,q)$ satisfies the equation

$$\cosh^2 \frac{d_\Gamma(p,q)}{2k_\Gamma} = \frac{f_{pq}^2}{f_{pp}f_{qq}}.$$

PROOF. Given points p and q, the isotropic points x and y collinear with them have parameters (λ, μ), relative to p and q, satisfying

$$(pAp^{\mathrm{T}})\lambda^2 + 2(pAq^{\mathrm{T}})\lambda\mu + (qAq^{\mathrm{T}})\mu^2 = 0,$$

or

$$f_{pp}\lambda^2 + 2f_{pq}\lambda\mu + f_{qq}\mu^2 = 0.$$

(This is by Lemma 4.3.1.) Solving this equation for λ in terms of μ, we get

$$\lambda = \frac{-f_{pq} \pm \sqrt{f_{pq}2 - f_{pp}f_{qq}}}{f_{pp}}\mu.$$

We therefore assign parameters to x, y, p, and q as follows:

$$\begin{aligned} x &= \left(-f_{pq} + \sqrt{f_{pq}^2 - f_{pp}f_{qq}}, f_{pp}\right), \\ y &= \left(-f_{pq} - \sqrt{f_{pq}^2 - f_{pp}f_{qq}}, f_{pp}\right), \\ p &= (1,0), \\ q &= (0,1). \end{aligned}$$

The cross ratio is then easily seen to be the given expression.

Now $d_\Gamma(p,q) = k_\Gamma \ln R(xypq)$, and $\frac{1}{2}\ln R(xypq) = \ln\sqrt{R(xypq)}$, so

$$\begin{aligned} \cosh^2 \frac{d_\Gamma(p,q)}{2k_\Gamma} &= \left[\tfrac{1}{2}\left(e^{\ln\sqrt{R(xypq)}} + e^{-\ln\sqrt{R(xypq)}}\right)\right]^2 \\ &= \left[\frac{1}{2}\left(\sqrt{R(xypq)} + \frac{1}{\sqrt{R(xypq)}}\right)\right]^2 \\ &= \frac{1}{4}\left[R(xypq) + 2 + \frac{1}{R(xypq)}\right]. \end{aligned}$$

Now putting in the expression for $R(xypq)$, we find after a straightforward but tedious simplification that

$$\cosh^2 \frac{d_\Gamma(p,q)}{2k_\Gamma} = \frac{f_{pq}^2}{f_{pp}f_{qq}}. \qquad \square$$

The dual of Theorem 3 gives corresponding results for angles, as follows:

Theorem 4. *If L and M are nonisotropic lines, concurrent with the isotropic lines X and Y, then*

$$R(XYLM) = \frac{-F_{LM} - \sqrt{F_{LM}^2 - F_{LL}F_{MM}}}{-F_{LM} + \sqrt{F_{LM}^2 - F_{LL}F_{MM}}},$$

and $D_\Gamma(L, M)$ *satisfies the equation*

$$\cosh^2 \frac{D_\Gamma(L, M)}{2K_\Gamma} = \frac{F_{LM}^2}{F_{LL}F_{MM}}.$$

Points that are conjugate relative to the absolute Γ, that is, that lie on each other's polars, have an interesting distance property, as the next theorem states.

Theorem 5. *Nonisotropic points* p *and* q *are conjugate relative to* Γ *if and only if*

$$d_\Gamma(p, q) = (2n + 1)k_\Gamma \pi i.$$

This theorem is proved using Theorem 4.6.3, Corollary 3, or Exercise 4.3.17.

Exercises 5.1

1. Let $z = x + iy$ be a complex number. If $|z| = \sqrt{x^2 + y^2}$ and $e^{i\theta} = \cos\theta + i\sin\theta$, show that $z = |z|e^{i\theta}$ with $\cos\theta = x/|z|$ and $\sin\theta = y/|z|$, and then show that $\ln z = \ln|z| + i\theta$.

2. Find (a) $\ln i$, (b) $\ln(1/i)$, (c) $\ln(-2)$.

3. Using $\cosh z = \frac{1}{2}(e^z + e^{-z})$ and $\sinh z = \frac{1}{2}(e^z - e^{-z})$, show that

   (a) $\cosh iz = \cos z$, **(b)** $\sinh iz = i\sin z$,
   (c) $\cos z = \frac{1}{2}(e^{iz} + e^{-iz})$, **(d)** $\sin z = (1/2i)(e^{iz} - e^{-iz})$,
   (e) $\cos iz = \cosh z$, **(f)** $\sin iz = i\sinh z$.

4. State the dual of Theorem 2.

5. Carry out the missing calculations in the proof of Theorem 3.

6. Prove Theorem 5.

7. State the dual of Theorem 5.

8. If p is a point and L is a line not on p, show that some point on L is conjugate to p. Dualize this result.

9. Given a nonisotropic point p and a nonisotropic line L, not on p, let M be the line on p conjugate to L, and let $q = L \cap M$. We define $d_\Gamma(p, q)$ to be the *distance from p to L*. Show that the distance d from p to L satisfies the equation

$$\cosh^2 \frac{d}{2k_\Gamma} = 1 - \frac{(pL^{\mathrm{T}})^2}{f_{PP}F_{LL}}.$$

10. Find the distance from a nonisotropic point to its polar.

11. By dualizing Exercise 9, define the *angle from L to p*, and compare it with the distance from p to L.

Section 5.2. The Triangle in π_C

In this section, we assume a metric (d_Γ, D_Γ) on π_C, the absolute having point conic $xAx^T = 0$ and line conic $LA^{-1}L^T = 0$. We let $p_i = [p_{i1}, p_{i2}, p_{i3}]$, $i = 1,2,3$, be vertices of a triangle, and let $S_1 = p_2p_3$, $S_2 = p_3p_1$, $S_3 = p_1p_2$ be the sides of the triangle. Let side S_i have coordinates $\langle s_{i1}, s_{i2}, s_{i3}\rangle$, $i = 1,2,3$.

To obtain some analytic identities for this triangle, we let

$$f_{ij} = p_iAp_j^T,$$

and

$$F_{ij} = S_iA^{-1}S_j^T,$$

and we construct the matrices

$$P = (p_{ij}), \quad S = (s_{ij}), \quad f = (f_{ij}), \quad F = (F_{ij}).$$

In what follows, i, j, k will always be a cyclic permutation of $1,2,3$. Thus, for example, S_i is the line p_jp_k.

Theorem 1. *S is the cofactor matrix of P, so that $S = |P|(P^{-1})^T$.*

PROOF. Since $S_i = p_jp_k$, we have

$$\langle s_{i1}, s_{i2}, s_{i3}\rangle = \left\langle \begin{vmatrix} p_{j2} & p_{j3} \\ p_{k2} & p_{k3} \end{vmatrix}, \begin{vmatrix} p_{j3} & p_{j1} \\ p_{k3} & p_{k1} \end{vmatrix}, \begin{vmatrix} p_{j1} & p_{j2} \\ p_{k1} & p_{k2} \end{vmatrix} \right\rangle,$$

so that s_{ij} is the cofactor of p_{ij} in P. Hence S is the cofactor matrix of P. Since the inverse P^{-1} of P exists (p_1, p_2, p_3 are not collinear) and is

$$P^{-1} = \frac{1}{|P|}(\text{cofactor matrix of } P)^T,$$

it follows that $S = |P|(P^{-1})^T$. □

Theorem 2. *$f = PAP^T$ and $F = SA^{-1}S^T$.*

PROOF. Since $f_{ij} = p_iap_j^T$, the first equation follows immediately. The second is seen the same way. □

Theorem 3. *The matrices f and F are symmetric and nonsingular.*

PROOF. Since P and A are nonsingular, the product $f = PAP^T$ is nonsingular. Also,

$$f^T = (PAP^T)^T = P^{TT}A^TP^T = PAP^T = f$$

because A is symmetric, so f is symmetric. Similarly for F. □

Theorem 4. $|f|f^{-1}=|A|F$.

PROOF. First, we have

$$|f|=|PAP^{\mathrm{T}}|=|P||A||P^{\mathrm{T}}|=|P|^2|A|.$$

Second we have

$$\begin{aligned} f^{-1}=(PAP^{\mathrm{T}})^{-1}&=(P^{\mathrm{T}})^{-1}A^{-1}P^{-1}\\ &=(P^{-1})^{\mathrm{T}}A^{-1}(P^{-1}). \end{aligned}$$

Now by Theorem 1,

$$(P^{-1})^{\mathrm{T}}=\frac{1}{|P|}S \quad \text{and} \quad P^{-1}=\frac{1}{|P|}S^{\mathrm{T}};$$

hence

$$\begin{aligned} f^{-1}&=\left(\frac{1}{|P|}S\right)A^{-1}\left(\frac{1}{|P|}S^{\mathrm{T}}\right)\\ &=\frac{1}{|P|^2}SA^{-1}S^{\mathrm{T}}\\ &=\frac{1}{|P|^2}F. \end{aligned}$$

Putting these together, we get

$$|f|f^{-1}=|P|^2|A|\cdot\frac{1}{|P|^2}F=|A|F.$$

□

Theorem 5.

$$F_{ii}=\frac{f_{jj}f_{kk}-f_{jk}^2}{|A|}$$

and

$$F_{ij}=\frac{f_{ik}f_{jk}-f_{ij}f_{kk}}{|A|}.$$

PROOF. We use the fact that the inverse f^{-1} of f is $1/|f|$ multiplied by the transposed cofactor matrix of f; that is,

$$\begin{aligned} F&=\frac{|f|}{|A|}\cdot\frac{1}{|f|}(\text{cof. matr. of } f)^{\mathrm{T}}\\ &=\frac{1}{|A|}(\text{cof. matr. of } f)^{\mathrm{T}}. \end{aligned}$$

Writing out f,

$$f=\begin{pmatrix} f_{11} & f_{12} & f_{13}\\ f_{21} & f_{22} & f_{23}\\ f_{31} & f_{32} & f_{33} \end{pmatrix},$$

we can read off the cofactors. Thus

$$F_{ii} = \frac{1}{|A|}(f_{jj}f_{kk} - f_{jk}f_{kj}).$$

But f is symmetric so $f_{kj} = f_{jk}$, and

$$F_{ii} = \frac{1}{|A|}(f_{jj}f_{kk} - f_{jk}^2).$$

Similarly,

$$F_{ij} = \frac{1}{|A|}(f_{ik}f_{jk} - f_{ij}f_{kk}).$$

□

Theorem 6.

$$f_{ii} = \frac{|A|^2}{|f|}(F_{jj}F_{kk} - F_{jk}^2)$$

and

$$f_{ij} = \frac{|A|^2}{|f|}(F_{ik}F_{jk} - F_{ij}F_{kk}).$$

PROOF. Using Theorem 4, we can get

$$f = \frac{|A|^2}{|f|}\,(\text{cof. matr. of } F)^{\mathrm{T}},$$

and then we proceed as in the proof of Theorem 5. □

Exercises 5.2

1. Show that $d_\Gamma(p_i, p_j)$ satisfies

$$\cosh^2 \frac{d_\Gamma(p_i, p_j)}{2k_\Gamma} = \frac{f_{ij}^2}{f_{ii}f_{jj}}.$$

2. Show that $D_\Gamma(S_i, S_j)$ satisfies

$$\cosh^2 \frac{D_\Gamma(S_i, S_j)}{2K_\Gamma} = \frac{F_{ij}^2}{F_{ii}F_{jj}}.$$

3. Let h_i be the altitude of $\triangle p_1p_2p_3$ from vertex p_i, $i = 1, 2, 3$. Show that h_i satisfies

$$\cosh^2 \frac{h_i}{2k_\Gamma} = 1 - \frac{|P|^2}{f_{ii}F_{ii}}.$$

Section 5.3. Elliptic Geometry

With a metric on π_C, we can now specialize in such a way as to obtain an analytic version of elliptic geometry, first described by G. F. B. Riemann (1826–1866) in 1851. To do so, we choose a particular absolute, and then study only certain points and lines in π_C.

Recall that the real projective plane π_R is a subplane of π_C. Points of π_R are called *real* points in π_C. A point conic $\Gamma: xAx^{\mathrm{T}} = 0$ is called a *real* conic in case its matrix A is real. A real conic may or may not contain real points. If point conic $\Gamma: xAx^{\mathrm{T}} = 0$ is real, then so is its derived line conic $\Gamma' : LA^{-1}L^{\mathrm{T}} = 0$. It is shown in the exercises that if a real point conic Γ contains no real points, then its derived conic Γ' contains no real lines.

Definition 1. An *elliptic geometry* is π_R as a subplane of π_C, under a metric whose absolute is real and contains no real points.

To avoid some unnecessary difficulties, we choose as the absolute the point conic $\Gamma: x_1^2 + x_2^2 + x_3^2 = 0$, which is real and contains no real points. Its matrix is the identity matrix, and its derived conic has equation $l_1^2 + l_2^2 + l_3^2 = 0$. We also choose the scale constants to be

$$k_\Gamma = K_\Gamma = \frac{1}{2i}\,.$$

Now we shall specialize the results of the previous section for a triangle $p_1p_2p_3$. In the notation of Section 5.2, we have the following theorems.

Theorem 1. *$f_{ii}f_{jj} - f_{ij}^2 > 0$ and $F_{ii}F_{jj} - F_{ij}^2 > 0$.*

PROOF. Since A is the identity matrix, we have

$$f_{ij} = p_iAp_j^{\mathrm{T}} = p_ip_j^{\mathrm{T}} = p_{i1}p_{j1} + p_{i2}p_{j2} + p_{i3}p_{j3}.$$

Hence

$$\begin{aligned} f_{ii}f_{jj} - f_{ij}^2 &= \left(p_{i1}^2 + p_{i2}^2 + p_{i3}^2\right)\left(p_{j1}^2 + p_{j2}^2 + p_{j3}^2\right) \\ &\quad - \left(p_{i1}p_{j1} + p_{i2}p_{j2} + p_{i3}p_{j3}\right)^2 \\ &= \left(p_{i1}^2p_{j2}^2 - 2p_{i1}p_{j2}p_{i2}p_{j1} + p_{i2}^2p_{j1}^2\right) \\ &\quad + \left(p_{i1}^2p_{j3}^2 - 2p_{i1}p_{j3}p_{i}p_{j1} + p_{i3}^2p_{j1}^2\right) \\ &\quad + \left(p_{i2}^2p_{j3}^2 - 2p_{i2}p_{j3}p_{i3}p_{j2} + p_{i3}^2p_{j2}^3\right) \\ &= \left(p_{i1}p_{j2} - p_{i2}p_{ji}\right)^2 + \left(p_ip_{j3} - p_{i3}p_{j1}\right)^2 + \left(p_{i2}p_{i3} - p_{i3}p_{j2}\right)^2 \\ &= s_{k3}^2 + s_{k2}^2 + s_{k1}^2 > 0, \end{aligned}$$

since not all the coordinates of S_k are zero. Similarly for $F_{ii}F_{jj} - F_{ij}^2$. □

Corollary. $-1 < f_{ij}/\sqrt{f_{ii}f_{jj}} < 1$ *and* $-1 < F_{ij}/\sqrt{F_{ii}F_{jj}} < 1$.

PROOF. Since $f_{ii}f_{jj} - f_{ij}^2 > 0$, we have

$$0 \leqslant f_{ij}^2 < f_{ii}f_{jj},$$

$$|f_{ij}| < \sqrt{f_{ii}f_{jj}},$$

$$-\sqrt{f_{ii}f_{jj}} < f_{ij} < \sqrt{f_{ii}f_{jj}},$$

$$-1 < \frac{f_{ij}}{\sqrt{f_{ii}f_{jj}}} < 1.$$

The other inequality is similarly proved. □

Because of the corollary, we may choose real numbers θ_{ij}, $0 < \theta_{ij} < \pi$, and ϕ_{ij}, $0 < \phi_{ij} < \pi$, such that

$$\cos\theta_{ij} = \frac{f_{ij}}{\sqrt{f_{ii}f_{jj}}} \quad \text{and} \quad \cos\phi_{ij} = \frac{F_{ij}}{\sqrt{F_{ii}F_{jj}}}.$$

Doing so, we get the following theorems.

Theorem 2. $d_\Gamma(p_i, p_j) = \theta_{ij} + n\pi$, *n an integer.*

PROOF. From Theorem 5.1.3, we have

$$\begin{aligned}
R(xyp_ip_j) &= \frac{-f_{ij} - \sqrt{f_{ij}^2 - f_{ii}f_{jj}}}{-f_{ij} + \sqrt{f_{ij}^2 - f_{ii}f_{jj}}} \\
&= \frac{-\left(f_{ij}/\sqrt{f_{ii}f_{jj}}\right) - \sqrt{(f_{ij}^2/f_{ii}f_{jj}) - 1}}{-\left(f_{ij}/\sqrt{f_{ii}f_{jj}}\right) + \sqrt{(f_{ij}^2/f_{ii}f_{jj}) - 1}} \\
&= \frac{-\cos\theta_{ij} - \sqrt{\cos^2\theta_{ij} - 1}}{-\cos\theta_{ij} + \sqrt{\cos^2\theta_{ij} - 1}} \\
&= \frac{-\cos\theta_{ij} - i\sin\theta_{ij}}{-\cos\theta + i\sin\theta_{ij}} \\
&= \frac{\cos\theta_{ij} + i\sin\theta_{ij}}{\cos\theta_{ij} - i\sin\theta_{ij}} \\
&= \frac{e^{i\theta_{ij}}}{e^{-i\theta_{ij}}} = e^{2i\theta_{ij}}.
\end{aligned}$$

Hence

$$\begin{aligned} d_\Gamma(p_i, p_j) &= k_\Gamma \ln P_x(xyp_ip_j) \\ &= k_\Gamma \ln e^{i(2\theta_{ij})} \\ &= \frac{1}{2i}[0 + i(2\theta_{ij} + 2n\pi)] \\ &= \theta_{ij} + n\pi. \end{aligned}$$

□

Similarly, we have the following theorem for angles.

Theorem 3. *$D_\Gamma(S_i, S_j) = \phi_{ij} + n\pi$, n an integer.*

Simplifying our notation a little more, we let

$$d_i = d_\Gamma(p_j, p_k),$$
$$\alpha_i = D_\Gamma(p_ip_j, p_ip_k) = D_\Gamma(S_k, S_j).$$

Then we can write the following theorem.

Theorem 4.

1.

$$\cos d_i = \frac{f_{jk}}{\sqrt{f_{jj}f_{kk}}},$$

2.

$$\sin d_i = \sqrt{\frac{f_{jj}f_{kk} - f_{jk}^2}{f_{jj}f_{kk}}} = \sqrt{\frac{F_{ii}}{f_{jj}f_{kk}}},$$

3.

$$\cos \alpha_i = \frac{-F_{jk}}{\sqrt{F_{jj}F_{kk}}},$$

4.

$$\sin \alpha_i = \sqrt{\frac{F_{jj}F_{kk} - F_{jk}^2}{F_{jj}F_{kk}}} = \sqrt{\frac{|f| f_{ii}}{F_{jj}F_{kk}}}.$$

PROOF. For part 1, we use Theorem 5.1.3:

$$\begin{aligned} \cosh^2 \frac{d_i}{2k_\Gamma} &= \cosh^2 \frac{d_\Gamma(p_jp_k)}{2k_\Gamma} \\ &= \frac{f_{jk}^2}{f_{jj}f_{kk}}. \end{aligned}$$

But also,

$$\frac{d_i}{2k_\Gamma} = \frac{d_i}{1/i} = id_i,$$

and from Exercise 5.1.3, we have

$$\cosh^2 id_i = \cos^2 d_i.$$

The result follows. Proof of the other parts is similar. □

Now we can state the rules of trigonometry for elliptic geometry.

Theorem 5. The Laws of Cosines.

1. $\cos d_i = \cos d_j \cos d_k + \sin d_j \sin d_k \cos \alpha_i,$
2. $\cos \alpha_i = -\cos \alpha_j \cos \alpha_k + \sin \alpha_j \sin \alpha_k \cos d_i.$

Theorem 6. The Law of Sines.

$$\frac{\sin \alpha_1}{\sin d_1} = \frac{\sin \alpha_2}{\sin d_2} = \frac{\sin \alpha_3}{\sin d_3} = \sqrt{\frac{|f| f_{11} f_{22} f_{33}}{F_{11} F_{22} F_{33}}}.$$

These theorems are proved using the identities of Theorem 4. They are vaguely like the familiar Euclidean laws with the same names, but there are some striking differences. Note the duality present in the laws of cosines, not a feature of Euclidean geometry. In the law of sines, $\sin d_i$ appears instead of d_i.

With these laws in hand, we can develop some further properties of elliptic geometry. These also differ strikingly from the properties of familiar Euclidean geometry.

The first of these tells us that the angles of a triangle do not add up to π.

Theorem 7. The Angle-Sum Theorem. *If $\alpha_1, \alpha_2, \alpha_3$ are angles of a triangle in elliptic geometry and $0 < \alpha_1 \leqslant \alpha_2 \leqslant \alpha_3 < \pi$, then $\alpha_1 + \alpha_2 + \alpha_3 > \pi$.*

PROOF. If $0 < \alpha_1 \leqslant \alpha_2 \leqslant \alpha_3 < \pi$, then $-\pi < -\alpha_1 - \alpha_2 < \alpha_3 - \alpha_1 - \alpha_2 < \pi - \alpha_1 - \alpha_2 < \pi$. Then by the second law of cosines,

$$\cos \alpha_3 = -\cos \alpha_1 \cos \alpha_2 + \sin \alpha_1 \sin \alpha_2 \cos d_3.$$

The cosine addition formula tells us

$$\cos(\alpha_1 + \alpha_2) = \cos \alpha_1 \cos \alpha_2 - \sin \alpha_1 \sin \alpha_2.$$

Adding these two expressions, we get

$$\cos \alpha_3 + \cos(\alpha_1 + \alpha_2) = \sin \alpha_1 \sin \alpha_2 (\cos d_3 - 1) < 0.$$

From the trigonometric identity

$$\cos x + \cos y = 2 \cos \tfrac{1}{2}(x + y) \cos \tfrac{1}{2}(x - y),$$

we can now write

$$2 \cos \tfrac{1}{2}(\alpha_1 + \alpha_2 + \alpha_3) \cos \tfrac{1}{2}(\alpha_3 - \alpha_1 - \alpha_2) = \sin \alpha_1 \sin \alpha_2 (\cos d_3 - 1).$$

Now $-\frac{1}{2}\pi < \frac{1}{2}(\alpha_3 - \alpha_1 - \alpha_2) < \frac{1}{2}\pi$, so $\cos\frac{1}{2}(\alpha_3 - \alpha_1 - \alpha_2) > 0$. Therefore $\cos\frac{1}{2}(\alpha_1 + \alpha_2 + \alpha_3) < 0$, so that $\frac{1}{2}(\alpha_1 + \alpha_2 + \alpha_3) > \frac{1}{2}\pi$, or $\alpha_1 + \alpha_2 + \alpha_3 > \pi$. □

The second result tells us that similar triangles of different sizes do not exist.

Theorem 8. *If two triangles are similar in elliptic geometry, then they are congruent.*

PROOF. By the second law of cosines, we can find the sides in terms of the angles, so once the angles are given, the sides are determined. □

The fact that an AAA congruence theorem exists says that the area of a triangle can be computed from the angles. The formula is very simple, though its derivation is not; so we state the formula without proof.

Theorem 9. *The area of* $\triangle p_1p_2p_3$ *in elliptic geometry is*

$$\mathcal{A} = \alpha_1 + \alpha_2 + \alpha_3 - \pi.$$

Since the area formula is in the context of choosing $\alpha_i < \pi$, that implies that a triangle cannot have area more than 2π. Thus arbitrarily large triangles do not exist. That in turn implies (though we shall not show it) that the entire elliptic plane is finite in area.

The trigonometric laws for elliptic geometry turn out to be identical to the formulas of spherical trigonometry on a sphere of radius 1. Thus elliptic geometry is qualitatively the same as the geometry on the surface of a sphere.

Exercises 5.3

1. Derive part 2 of Theorem 4.

2. Prove the first law of cosines.

3. Prove the second law of cosines.

4. Prove the law of sines.

5. For $p_1 = [1, 1, 0]$, $p_2 = [0, 2, 1]$, and $p_3 = [1, -1, 2]$, find the sides and angles of $\triangle p_1p_2p_3$.

6. Let α and β be the angles opposite sides a and b in a right triangle with hypotenuse c. Derive the following formulas:

   (a) $\sin b = \sin c \sin \beta$,
   (b) $\cos c = \cos a \cos b$,
   (c) $\cos \alpha = \cos a \sin \beta$,
   (d) $\cos \alpha \cos \beta = \sin \alpha \sin \beta \cos c$.

7. Suppose $\triangle p_1 p_2 p_3$ is very small. In the laws of cosines and sines, replace $\cos d_i$ and $\sin d_i$ by their Maclauren series and neglect third- and higher-order terms, thus showing that the first law of cosines becomes the familiar law of cosines, the law of sines becomes the familiar law of sines, and the second law of cosines becomes the Euclidean angle-sum theorem. That is, "in the small," elliptic geometry becomes Euclidean.

Section 5.4. Hyperbolic Geometry

If we specialize the metric differently, a subset of π_C becomes the classical hyperbolic geometry, presented about 1830 independently by J. Bolyai (1802–1860) and N. I. Lobatchevski (1793–1856). Incidentally, the ideas of both men were anticipated by C. F. Gauss (1777–1855), who developed the same geometry by 1820 or so but never published it.

This time we let the absolute $\Gamma : xAx^{\mathrm{T}} = 0$ be a real conic that contains at least one real point, and has $|A| > 0$. We then call a real point p *interior* if $pAp^{\mathrm{T}} > 0$ and *exterior* if $pAp^{\mathrm{T}} < 0$. A real line L is *interior* if $LA^{-1}L^{\mathrm{T}} < 0$ and *exterior* if $LA^{-1}L^{\mathrm{T}} > 0$.

Definition 1. A *hyperbolic geometry* is the geometry consisting of the interior points and lines of π_R under a metric on π_C whose absolute is real and contains at least one real point.

To make life simple, we choose the absolute to have point conic Γ: $-x_1^2 - x_2^3 + x_3^2 = 0$. Its matrix has determinant 1, and the envelope of tangent lines has equation $\Gamma' : -l_1^2 - l_2^2 + l_3^2 = 0$. We specify the scale constants to be

$$k_\Gamma = \tfrac{1}{2} \quad \text{and} \quad K_\Gamma = \frac{1}{2i} .$$

We shall also use the same notation as previously.

Theorem 1. *If p_i and p_j are interior points, then $f_{ii}f_{jj} - f_{ij}^2 < 0$.*

PROOF. Since p_i and p_j are interior points,

$$f_{ii} = -p_{i1}^2 - p_{i2}^2 + p_{i3}^2 = p_i A p_i^{\mathrm{T}} > 0$$

and

$$f_{jj} = -p_{j1}^2 - p_{j2}^2 + p_{j3}^2 = p_j A p_j^{\mathrm{T}} > 0.$$

Therefore $p_{i3} \neq 0$ and $p_{j3} \neq 0$; assume $p_{i3} = p_{j3} = 1$. Then set $p_{i1} = r_i \cos A_i$ and $p_{i2} = r_i \sin A_i$, with $r_i > 0$. Then $r_i < 1$. It follows that

$$\begin{aligned} f_{ii} &= -r_i^2 \cos^2 A_i - r_i^2 \sin^2 A_i + 1 \\ &= 1 - r_i^2. \end{aligned}$$

Similarly,

$$f_{jj} = 1 - r_j^2.$$

Then

$$\begin{aligned} f_{ij} &= -p_{i1}p_{j1} - p_{i2}p_{j2} + p_{i3}p_{j3} \\ &= -r_ir_j\cos A_i\cos A_j - r_ir_j\sin A_i\sin A_j + 1 \\ &= 1 - r_ir_j\cos(A_i - A_j). \end{aligned}$$

It follows from the identity $\cos\theta = 1 - 2\sin^2\frac{1}{2}\theta$ that

$$\begin{aligned} f_{ij} &= 1 - r_ir_j\left[1 - 2\sin^2\tfrac{1}{2}(A_i - A_j)\right] \\ &= 1 - r_ir_j + 2r_ir_j\sin^2\tfrac{1}{2}(A_i - A_j). \end{aligned}$$

Then by a tedious but straightforward calculation, we find that

$$\begin{aligned} f_{ii}f_{jj} - f_{ij}^2 = -(r_i - r_j)^2 &- 4r_i^2r_j^2\sin^4\tfrac{1}{2}(A_i - A_j) \\ &- 4r_ir_j\sin^2\tfrac{1}{2}(A_i - A_j) < 0. \end{aligned}$$

□

Theorem 2. *If p_i and p_j are interior points and $p_{i3} > 0$ and $p_{j3} > 0$, then $f_{ij} > 0$.*

PROOF. From the proof of Theorem 1, we have $0 < r_i < 1$ and $0 < r_j < 1$. Hence $0 < r_ir_j < 1$, and

$$f_{ij} = 1 - r_ir_j + 2r_ir_j\sin^2\tfrac{1}{2}(A_i - A_j) > 0.$$

□

We shall always choose the coordinates of point p_i so that $p_{i3} > 0$. Then we have

Theorem 3. $f_{ij}\Big/\sqrt{f_{ii}f_{jj}} > 1$.

PROOF. By Theorem 1,

$$f_{ij}^2 > f_{ii}f_{jj},$$

or

$$\frac{f_{ij}^2}{f_{ii}f_{jj}} > 1.$$

Taking the square root and noting that $f_{jj} > 0$ by Theorem 2, we get the desired result. □

Theorem 4. *If S_i and S_j are two real lines meeting at the interior point p_k, then $F_{ii}F_{jj} - F_{ij}^2 > 0$.*

PROOF. A straightforward calculation shows that

$$\begin{aligned} F_{ii}F_{jj} - F_{ij}^2 &= (-s_{i1}^2 - s_{i2}^2 + s_{i3}^2)(-s_{ji}^2 - s_{j2}^2 + s_{j3}^2) \\ &\quad - (-s_{i1}s_{j1} - s_{i2}s_{j2} + s_{i3}s_{j3})^2 \\ &= -(s_{i2}s_{j3} - s_{i3}s_{j2})^2 - (s_{i3}s_{i1} - s_{i1}s_{j3})^2 + (s_{i1}s_{j2} - s_{j1}s_{i2})^2 \\ &= -p_{k1}^2 - p_{k2}^2 + p_{k3}^2 > 0 \end{aligned}$$

because p_k is an interior point. □

Corollary. $-1 < F_{ij}/\sqrt{F_{ii}F_{jj}} < 1$.

Theorem 5. *The line on two interior points is an interior line.*

PROOF. Let S_k be the line on interior points p_i and p_j. Then

$$S_k A^{-1} S_k^{\mathrm{T}} = F_{kk} = \frac{1}{|A|}\left(f_{ii}f_{jj} - f_{ij}^2\right) < 0$$

by Theorem 5.2.5 and Theorem 1. Thus S_k is an interior line. □

Theorem 6. *Any real line on an interior point is an interior line.*

PROOF. If p is an interior point, then we may write the coordinates of p as $[x, y, 1]$ with $x^2 + y^2 < 1$, and we may regard p as a point of α_R. If $L = \langle l_1, l_2, l_3 \rangle$ is any real line on p, then L is a line in α_R. The distance in α_R from the origin $[0, 0, 1]$ to L is

$$d = \frac{|l_3|}{\sqrt{l_1^2 + l_2^2}} .$$

But the distance to L does not exceed the distance to p from the origin, so

$$d \leqslant \sqrt{x^2 + y^2} < 1.$$

Therefore

$$\frac{|l_3|}{\sqrt{l_1^2 + l_2^2}} < 1,$$

or

$$|l_3| < \sqrt{l_1^2 + l_2^2}.$$

Hence

$$l_3^2 < l_1^2 + l_2^2,$$

or

$$-l_1^2 - l_2 + l_3^2 < 0,$$

and L is an interior line. □

Corollary. *Any real point on an exterior line is an exterior point.*

Now we are ready to do a little simplifying. Since $f_{ij}/\sqrt{f_{ii}f_{jj}} > 1$, we may choose a real number θ_{ij} such that

$$\cosh \theta_{ij} = \frac{f_{ij}}{\sqrt{f_{ii}f_{jj}}} .$$

By the Corollary to Theorem 4, we may choose a real number ϕ_{ij},

$0 < \phi_{ij} < \pi$, such that

$$\cos\phi_{ij} = \frac{-F_{ij}}{\sqrt{F_{ii}F_{jj}}}.$$

Then we have

Theorem 7.

$$d_\Gamma(p_i, p_j) = \theta_{ij} + n\pi i$$

and

$$D_\Gamma(S_i, S_j) = \phi_{ij} + n\pi.$$

The proof parallels the proof of Theorem 5.3.2, and will be omitted.

Theorem 8. *If* $d_i = d_\Gamma(p_j, p_k)$ *and* $\alpha_i = D_\Gamma(p_ip_j, p_ip_k) = D_\Gamma(S_k, S_j)$, *then*

1.

$$\cosh d_i = \frac{f_{jk}}{\sqrt{f_{jj}f_{kk}}},$$

2.

$$\sinh d_i = \sqrt{\frac{-F_{ii}}{f_{jj}f_{kk}}},$$

3.

$$\cos\alpha_i = \frac{F_{jk}}{\sqrt{F_{jj}F_{kk}}},$$

4.

$$\sin\alpha_i = \sqrt{\frac{|f|f_{ii}}{F_{jj}F_{kk}}}.$$

Again, the proof is omitted but is similar to that of Theorem 5.3.4. Now we get the trigonometric laws for the hyperbolic plane.

Theorem 9. The Laws of Cosines.

1. $\cosh d_i = \cosh d_j \cosh d_k - \sinh d_j \sinh d_k \cos\alpha_i$.
2. $\cos\alpha_i = -\cos\alpha_j \cos\alpha_k + \sin\alpha_j \sin\alpha_k \cosh d_i$.

Theorem 10. The Law of Sines.

$$\frac{\sin \alpha_1}{\sinh d_1} = \frac{\sin \alpha_2}{\sinh d_2} = \frac{\sin \alpha_3}{\sinh d_3} = \sqrt{\frac{|f| f_{11} f_{22} f_{33}}{F_{11} F_{22} F_{33}}} .$$

You should compare these with the corresponding laws for elliptic geometry and with the usual Euclidean laws, noting similarities and differences.

Further properties of hyperbolic geometry include the following.

Theorem 11. The Angle-Sum Theorem. *If $\alpha_1, \alpha_2, \alpha_3$ are angles of a triangle in hyperbolic geometry and $0 < \alpha_1 \leqslant \alpha_2 \leqslant \alpha_3 < \pi$, then $\alpha_1 + \alpha_2 + \alpha_3 < \pi$.*

Theorem 12. *The area of $\triangle p_1 p_2 p_3$ in hyperbolic geometry is*

$$\mathcal{A} = \pi - (\alpha_1 + \alpha_2 + \alpha_3).$$

Theorem 13. *In hyperbolic geometry, similar triangles are congruent.*

Hyperbolic geometry also can be regarded as the geometry on a certain surface, called the *pseudosphere*, whose equation in three dimensions is

$$z = \operatorname{sech}^{-1}\sqrt{x^2 + y^2} - \sqrt{1 - x^2 - y^2} .$$

This discovery was made by E. Beltrami (1835–1900) in 1868. The pseudosphere (see Figure 5.1) is a surface of constant negative curvature.

Exercises 5.4

1. Prove the Corollary to Theorem 6.

2. Prove Theorem 7.

3. Prove Theorem 8.

Figure 5.1. The pseudosphere

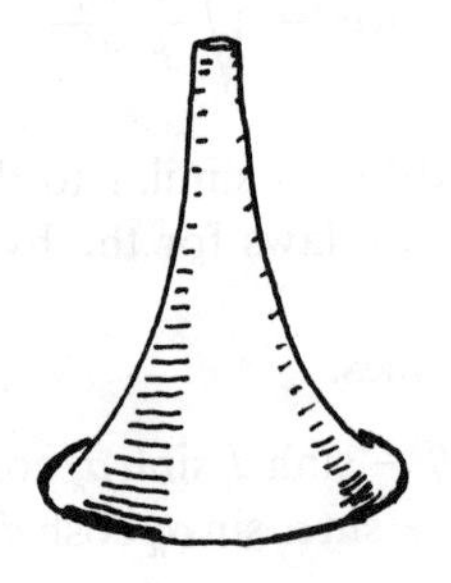

4. Prove the laws of cosines.

5. Prove the law of sines.

6. For $p_1 = [0,0,1]$, $p_2=[0,2,3]$, and $p_3 = [1,-1,2]$, find the sides and angles of $\triangle p_1 p_2 p_3$.

7. Prove the angle-sum theorem.

8. Let α and β be the angles opposite sides a and b of a right triangle with hypotenuse c. Derive the formulas:

(a) $\sinh b = \sinh c \sin \beta$,
(b) $\cosh c = \cosh a \cosh b$,
(c) $\cos \alpha = \cosh a \sin \beta$,
(d) $\cos \alpha \cos \beta = \sin \alpha \sin \beta \cosh c$.

9. Show that "in the small" hyperbolic geometry becomes Euclidean geometry.

Section 5.5. Euclidean Geometry

By suitably specializing the metric on π_C, we can get Euclidean geometry also. The idea is to let the absolute consist of the ideal line, and study real points not on it. But a singular absolute causes problems with the metric.

An ingenious way around this problem is to let the absolute consist of a family of conics, all but one of which is nonsingular, and then choose the singular one after all the calculations are done. This is carried out by a limit process.

Definition 1. A *Euclidean geometry* is the geometry of real points and lines in π_C not lying on the absolute, under a metric whose absolute, in the limit, consists of one real range and two complex pencils.

To be specific, we let the absolute be defined by the point conic

$$\Gamma : x(A + \lambda B)x^{\mathrm{T}} = 0$$

with

$$A = \begin{bmatrix} 0 & 0 & 0 \\ 0 & 0 & 0 \\ 0 & 0 & 1 \end{bmatrix} \quad \text{and} \quad B = \begin{bmatrix} 1 & 0 & 0 \\ 0 & 1 & 0 \\ 0 & 0 & -1 \end{bmatrix}.$$

We let $k_\Gamma = 1/2i\sqrt{\lambda}$ and $K_\Gamma = 1/2i$. The limit process is to let $\lambda \to 0^+$.

We let $C = A + B$, and use the notation of the previous sections.

Our first action is to worry about the envelope of tangents to Γ.

Theorem 1. *As $\lambda \to 0^+$, Γ becomes $xAx^{\mathrm{T}} = 0$ and Γ' becomes $LCL^{\mathrm{T}} = 0$.*

PROOF. That Γ becomes $xAx^{\mathrm{T}} = 0$ is obvious. To find Γ', we note that

$$A + \lambda B = \begin{pmatrix} \lambda & 0 & 0 \\ 0 & \lambda & 0 \\ 0 & 0 & 1-\lambda \end{pmatrix}$$

so that

$$(A + \lambda B)^{-1} = \begin{bmatrix} 1/\lambda & 0 & 0 \\ 0 & 1/\lambda & 0 \\ 0 & 0 & 1/(1-\lambda) \end{bmatrix}.$$

Since only proportionality counts, we may take instead of $(A + \lambda B)^{-1}$ the matrix

$$\lambda(1-\lambda)(A + \lambda B)^{-1} = \begin{pmatrix} 1-\lambda & 0 & 0 \\ 0 & 1-\lambda & 0 \\ 0 & 0 & \lambda \end{pmatrix}.$$

Then, as $\lambda \to 0$, this matrix becomes C. □

Now we begin dealing with distances and angles. First we present a theorem concerning distances.

Theorem 2. *As* $\lambda \to 0^+$,

$$-\frac{1}{\lambda}\sinh^2\frac{d_\Gamma(p_i, p_j)}{2k_\Gamma} \to d_\Gamma(p_i, p_j)^2.$$

PROOF. Using $k_\Gamma = 1/2i\sqrt{\lambda}$, we get

$$\begin{aligned} -\frac{1}{\lambda}\sinh^2\frac{d_\Gamma(p_i, p_j)}{2k_\Gamma} &= -\frac{1}{\lambda}\sinh^2\left[i\sqrt{\lambda}\, d_\Gamma(p_i, p_j)\right] \\ &= d_\Gamma(p_i, p_j)^2\left[\frac{\sinh\left[i\sqrt{\lambda}\, d_\Gamma(p_i, p_j)\right]}{i\sqrt{\lambda}\, d_\Gamma(p_i, p_j)}\right]^2, \end{aligned}$$

which approaches $d_\Gamma(p_i, p_j)^2$ as $\lambda \to 0^+$, by L'Hôpital's rule. □

Theorem 3.

$$\begin{aligned} d_\Gamma(p_i, p_j)^2 &= \frac{p_{i3}^2\left(p_j B p_j^{\mathrm{T}}\right) - 2p_{i3}p_{j3}\left(p_i B p_j^{\mathrm{T}}\right) + p_{j3}^2\left(p_i B p_i^{\mathrm{T}}\right)}{p_{i3}^2 p_{j3}^2} \\ &= \frac{S_k C S_k^{\mathrm{T}}}{p_{i3}^2 p_{j3}^2}. \end{aligned}$$

PROOF. Using Exercise 5.2.1, we can write

$$-\sinh^2 \frac{d_\Gamma(p_i, p_j)}{2k_\Gamma} = 1 - \cosh^2 \frac{d_\Gamma(p_i, p_j)}{2k_\Gamma}$$

$$= 1 - \frac{\left[p_i(A + \lambda B)p_j^{\mathrm{T}}\right]^2}{\left[p_i(A + \lambda B)p_i^{\mathrm{T}}\right]\left[p_j(A + \lambda B)p_j^{\mathrm{T}}\right]}$$

$$= 1 - \frac{\left[p_iAp_j^{\mathrm{T}} + \lambda p_iBp_j^{\mathrm{T}}\right]^2}{\left[p_iAp_i^{\mathrm{T}} + \lambda p_iBp_i^{\mathrm{T}}\right]\left[p_jAp_j^{\mathrm{T}} + \lambda p_jBp_j^{\mathrm{T}}\right]}.$$

Now $p_iAp_j^{\mathrm{T}} = p_{i3}p_{j3}$, so that $(p_iAp_i^{\mathrm{T}})(p_jAp_j^{\mathrm{T}}) = (p_iAp_j^{\mathrm{T}})^2$. Then

$$-\sinh^2 \frac{d_\Gamma(p_i, p_j)}{2k_\Gamma} = 1 - \frac{\left(p_iAp_j^{\mathrm{T}}\right)^2 + 2\lambda\left(p_iAp_j^{\mathrm{T}}\right)\left(p_iBp_j^{\mathrm{T}}\right) + \lambda^2(\cdots)}{\left(p_iAp_i^{\mathrm{T}}\right)\left(p_jAp_j^{\mathrm{T}}\right) + \lambda\left[p_{i3}^2\left(p_jBp_j^{\mathrm{T}}\right) + p_{j3}^2\left(p_iBp_i^{\mathrm{T}}\right)\right] + \lambda^2(\cdots)}$$

$$= \frac{\lambda\left[p_{i3}^2\left(p_jBp_j^{\mathrm{T}}\right) + p_{j3}^2\left(p_iBp_i^{\mathrm{T}}\right) - 2p_{i3}p_{j3}\left(p_iBp_j^{\mathrm{T}}\right)\right] + \lambda^2(\cdots)}{p_{i3}^2p_{j3}^2 + \lambda(\cdots)}$$

so that

$$-\frac{1}{\lambda}\sinh^2 \frac{d_\Gamma(p_i, p_j)}{2k_\Gamma}$$

$$= \frac{p_{i3}^2\left(p_jBp_j^{\mathrm{T}}\right) - 2p_{i3}p_{j3}\left(p_iBp_j^{\mathrm{T}}\right) + p_{j3}^2\left(p_iBp_i^{\mathrm{T}}\right) + \lambda(\cdots)}{p_{i3}^2p_{j3}^2 + \lambda(\cdots)}.$$

Thus, as $\lambda \to 0^+$, we get the stated result. The last equality is shown by complete expansion of both sides, a tedious process with no tricks (or fun) in it. □

The formula in Theorem 3 does not look right, but it really is the usual Euclidean distance formula. To see that, we identify the point $p_i = [p_{i1}, p_{i2}, p_{i3}]$ with point $(p_{i1}/p_{i3}, p_{i2}/p_{i3})$ of α_R. Then the distance between p_i and p_j is the square root of

$$\left(\frac{p_{i1}}{p_{i3}} - \frac{p_{j1}}{p_{j3}}\right)^2 + \left(\frac{p_{i2}}{p_{i3}} - \frac{p_{j2}}{p_{j3}}\right)^2 = \frac{(p_{i1}p_{j3} - p_{j1}p_{i3})^2 + (p_{i2}p_{j3} - p_{j2}p_{i3})^2}{p_{i3}^2p_{j3}^2}$$

$$= \frac{S_{k2}^2 + S_{k1}^2}{p_{i3}^2p_{j3}^2}$$

$$= \frac{S_kCS_k^{\mathrm{T}}}{p_{i3}^2p_{j3}^2} = d_\Gamma(p_i, p_j)^2.$$

To find angles, we first prove

Theorem 4. $0 \leqslant (S_i C S_j^{\mathrm{T}})^2/(S_i C S_i^{\mathrm{T}})(S_j C S_j^{\mathrm{T}}) < 1$.

PROOF. We start with

$$(s_{i1}s_{j2} - s_{i2}s_{j1})^2 = p_{k3}^2 > 0$$

because p_k is not an ideal (isotropic) point. Then

$$s_{i1}^2 s_{j2}^2 - 2s_{i1}s_{j2}s_{i2}s_{j1} + s_{i2}^2 s_{j1}^2 > 0,$$

$$s_{i1}^2 s_{j2}^2 + s_{i2}^2 s_{j1}^2 > 2s_{i1}s_{j2}s_{i2}s_{j1},$$

$$s_{i1}^2 s_{j1}^2 + s_{i1}^2 s_{j2}^2 + s_{i2}^2 s_{j1}^2 + s_{i2}^2 s_{j2}^2 > s_{i1}^2 s_{j1}^2 + 2s_{i1}s_{j1}s_{i2}s_{j2} + s_{i2}^2 s_{j2}^2,$$

$$(s_{i1}^2 + s_{i2}^2)(s_{j1}^2 + s_{j2}^2) > (s_{i1}s_{j1} + s_{i2}s_{j2})^2,$$

$$(S_i C S_i^{\mathrm{T}})(S_j C S_j^{\mathrm{T}}) > (S_i C S_j^{\mathrm{T}})^2 \geqslant 0.$$

and the stated result follows. □

Now let θ be a real number, $0 \leqslant \theta \leqslant \pi$, such that

$$\cos\theta = \frac{S_i C S_j^{\mathrm{T}}}{\sqrt{(S_i C S_i^{\mathrm{T}})(S_j C S_j^{\mathrm{T}})}}.$$

Then, following the example of Theorem 5.3.2, we can prove

Theorem 5. $D_\Gamma(S_i, S_j) = \theta + n\pi$.

Exercises 5.5

1. Show that Theorem 5 gives the angle you would get from the usual Euclidean methods, using slopes of the lines.

2. Derive the Euclidean law of cosines from Theorems 3 and 5.

3. Derive the Euclidean law of sines.

4. Derive the Euclidean angle-sum theorem.

Chapter 6

Projective Spaces

Thus far, we have focused our attention on projective planes of various types. Now we wish to consider projective geometries of other dimensions.

We begin by defining projective spaces and studying their basic properties. After considering some algebraic examples, we shall conclude with a brief discussion of mappings between projective spaces.

Section 6.1. Projective Spaces

The language of incidence structures developed at the beginning of the book is used to define projective spaces.

Definition 1. A *projective space* is an incidence structure $S = (\mathcal{P}, \mathcal{L}, \mathcal{I})$ such that

S1. If p and q are two points, then there is exactly one line on both p and q.
S2. If L is a line, then there are at least three points on L.
S3. If a, b, c, d are four points such that lines ab and cd meet, then lines ac and bd also meet.

It is easy to see that a projective plane satisfies the axioms of a projective space. Thus any projective plane is a projective space. Other simple examples of projective spaces are a single line (together with whatever points are on it), a single point, and the *empty space*, having no points or lines at all.

We shall denote the empty space by $\emptyset$. We shall also regard projective spaces as sets of points. Certain subsets of a projective space are themselves projective spaces; these we give a special name.

Definition 2. A *subspace* of a projective space S is a subset S' of S such that if p and q are two points of S contained in S', then every point of line pq is contained in S'.

It can now easily be shown that a subspace of a projective space is a projective space. For any projective space S, the empty space will be regarded as a subspace of S.

The next result shows how subspaces may be built out of smaller subspaces.

Theorem 1. *Let S' be a subspace of S, and let p be a point of S not in S'. Then the set $S'' = \{x \in S \mid \text{for some } q \in S', x \text{ is on line } pq\}$ is a subspace of S.*

PROOF. We must show that if a and b are points of S'', then line ab lies in S''. Consider the following cases.

Case 1. If $a, b \in S'$, then $ab \subseteq S' \subseteq S''$ by definition.

Case 2. If $a \in S'$ and $b = p$, then $ab \subseteq S''$ by definition.

Figure 6.1

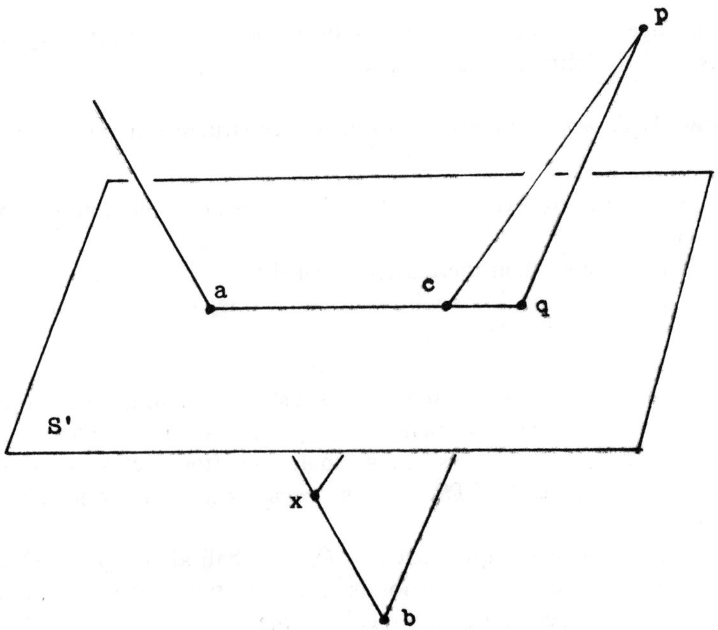

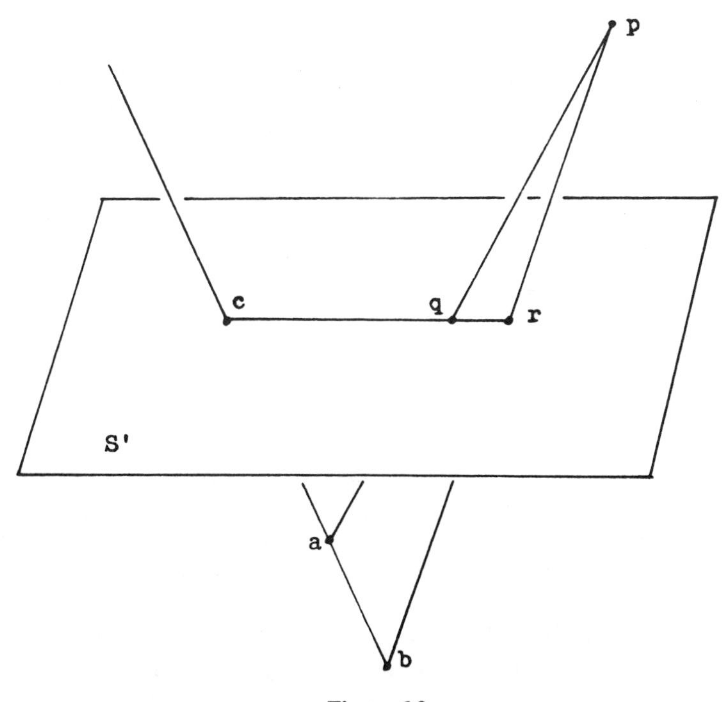

Figure 6.2

Case 3. If $a \in S'$ and $b \neq p$, $b \notin S'$, let b be collinear with p and q, $q \in S'$ (Figure 6.1). Let x be any other point of line ab. Since lines ax and pq meet at b, it follows from S3 that aq and xp meet at some point c. Since $aq \subseteq S'$, we have $c \in S'$, and therefore $x \in S''$. Thus $ab \subseteq S''$.

Case 4. If $a = p$ and $b \notin S'$, then $ab \subseteq S''$ by definition.

Case 5. If $a, b \notin S' \cup \{p\}$, let $q, r \in S'$ be such that a is on pq and b is on pr (Figure 6.2). Since aq and br meet at p, ab and qr meet at some point c by S3. Since $qr \subseteq S'$, $c \in S'$. Hence $ab \subseteq S''$ by Case 3. □

We can construct some particular subspaces now, using Theorem 1. Given a projective space S, let $S_0 = p_0$ be a point in S. Proceeding inductively, suppose that a subspace S_{n-1} of S has been constructed. To construct an S_n, let p_n be a point of S not lying in the S_{n-1}, and let

$$S_n = \{x \in S \mid \text{for some } q \in S_{n-1}, x \text{ is on line } p_n q\}.$$

Then S_n is a subspace of S by Theorem 1.

It is fairly obvious that an S_1 is just a line in S. To show that an S_2 is a projective plane takes a little more effort.

Theorem 2. *An S_2 is a projective plane.*

PROOF. Let an S_2 be constructed from a line S_1 and a point p_2. By S1 and Theorem 1, two points of the S_2 determine a line. If q and r are two points on S_1, let s and t be third points on lines pq and pr; then $qrst$ is a four-point. It remains only to show that two lines in S_2 meet. Let L and M be two lines in S_2, and consider the following cases.

Case 1. If $L = S_1$ and M is on p_2, let a be any other point on M. Then $a \in M \subseteq S_2$, so $M = ap_2$ meets $S_1 = L$ by definition.

Case 2. If $L = S_1$ and M is not on p_2, let a and b be two points on M. Then p_2a and p_2b meet S_1 at q and r (Figure 6.3), so by S3, $M = ab$ meets $qr = L$.

Case 3. If L is on p_2 and $M \neq S_1$, let q be the point at which L meets S_1, as it must by Case 1 (Figure 6.4). Let a be a point on M, and let b be the point at which M meets S_1, as in Case 2. Let p_2a meet S_1 at r. Since p_2a meets bq at r, it follows from S3 that $M = ab$ and $L = p_2q$ must meet.

Case 4. If neither L nor M is on p_2 or equal to S_1, let L and M meet S_1 at a and b, respectively, as they must by Case 2 (Figure 6.5). Let c be a second point on L, and let d be a second point on M. Then cd meets $S_1 = ab$ by Case 2, so $L = ac$ meets $bd = M$ by S3. □

Figure 6.3

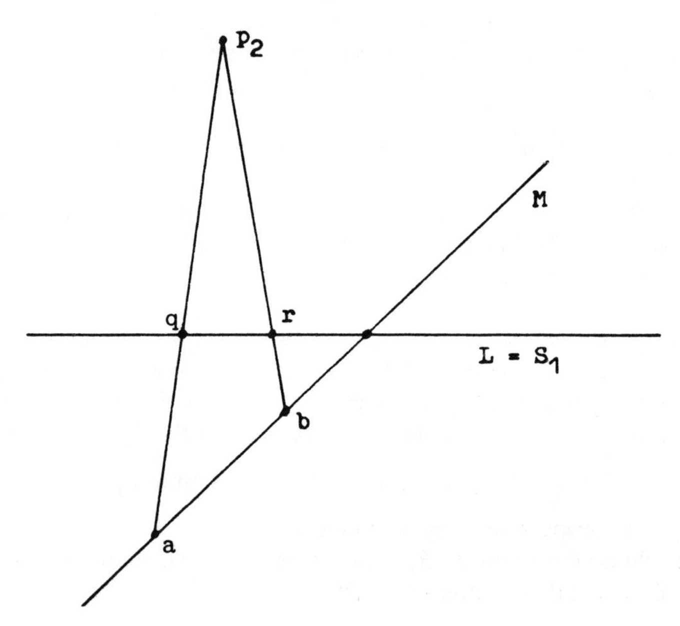

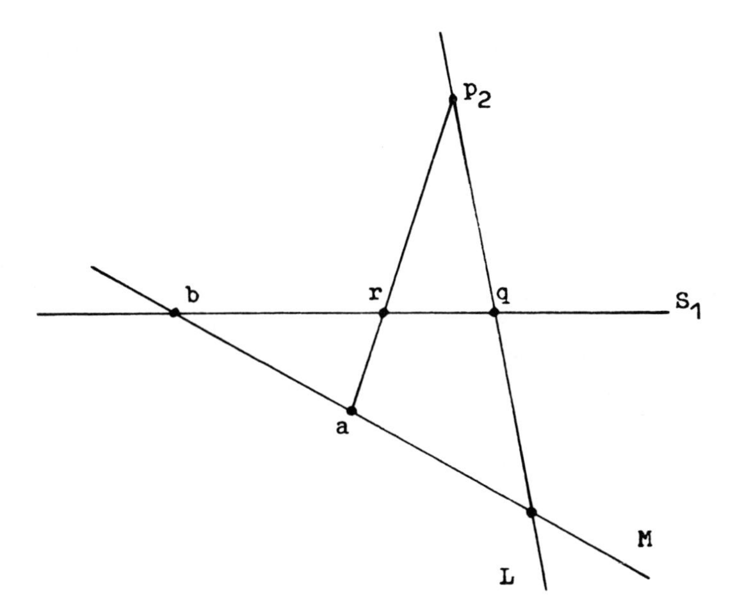

Figure 6.4

Figure 6.5

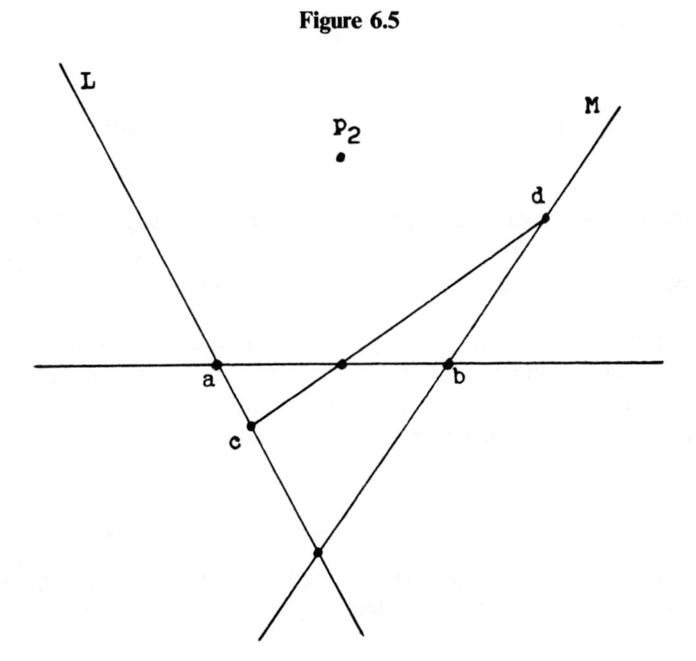

Another simple property of subspaces is stated in the following theorem.

Theorem 3. *The intersection of two subspaces of a projective space S is a subspace of S.*

The proof of this theorem is left as an exercise. The result can easily be extended to an arbitrary collection of subspaces. Thus if H is a subset of projective space S, the intersection of all subspaces containing H is a subspace containing H. This subspace is clearly the smallest subspace of S that contains H, and is called the *span* of H, denoted by $\langle H \rangle$. If $\langle H \rangle = S$, we call H a *spanning set* of S.

Some elementary properties of spans are found in the exercises. The following result is of particular use to us.

Lemma 1. *If H is a subset of the projective space S and p is a point of S such that $p \notin \langle H \rangle$, then*

$$\langle H \cup \{p\} \rangle = \{x \in S \mid \text{for some } q \in \langle H \rangle, x \text{ is on } pq\}.$$

PROOF. Let $T = \{x \in S \mid \text{for some } q \in \langle H \rangle, x \text{ is on } pq\}$. Then clearly $T \subseteq \langle H \cup \{p\} \rangle$. Conversely, let $x \in \langle H \cup \{p\} \rangle$. If $x \in \langle H \rangle$, then x is on px, so $x \in T$. If $x \notin \langle H \rangle$, then it is sufficient to show that px meets $\langle H \rangle$. But if not, then by Theorem 1 there is a subspace S' of S containing $\langle H \rangle$ and p, but not x, so that $x \notin \langle H \cup \{p\} \rangle$, a contradiction. Hence px meets $\langle H \rangle$, and $x \in T$. The lemma follows. □

Corollary. *If p is any point not lying in an S_{n-1}, then $\langle S_{n-1} \cup \{p\} \rangle$ is an S_n.*

To investigate projective spaces further, we need the concept of independence.

Definition 3. A subset H of a projective space is *independent* if and only if, for any point $p \in H$, $p \notin \langle H \setminus \{p\} \rangle$. A subset that is not independent is called *dependent*. If $p \in \langle H \rangle$, we say *p depends on H*.

For example, two points p and q in an S_2 are independent, for neither lies in the span of the other. A third point r does or does not depend on p and q according as r does or does not lie on line pq.

A useful result of independence is the following.

Lemma 2. *If $P_n = \{p_0, p_1, \ldots, p_n\}$ is a subset of a projective space S and no point of p_n depends on preceding points, then p_n is independent.*

PROOF. We shall proceed by induction on n. If $n = 1$, then $p_1 = \{p_0, p_1\}$. If p_1 does not depend on p_0, then $p_1 \neq p_0$. Hence p_0 does not depend on p_1, and P_1 is independent. Now assume that the result holds if $n = k$, and consider $n = k + 1$. Then in $P_{k+1} = \{p_0, p_1, \ldots, p_k, p_{k+1}\}$, no point depends on previous points. Suppose, for some p_i, that $p_i \in \langle P_{k+1} \backslash \{p_i\} \rangle$. Then, since p_{k+1} does not depend on preceding points, $i \neq k + 1$. Thus $p_i \notin \langle P_k \backslash \{p_i\} \rangle$ by the inductive hypothesis. Therefore, by Lemma 1, p_i and p_{k+1} are collinear with some point of $\langle P_k \backslash \{p_i\} \rangle$. But that implies, again by Lemma 1, that $p_{k+1} \in \langle P_k \rangle$, a contradiction. Thus P_{k+1} is independent, and the proof is complete. □

An independent spanning set of a projective space S is called a *basis* of S. For example, we have the following results.

Theorem 4. *An S_n has a basis of $n + 1$ points.*

PROOF. By the construction of an S_n, the set $\{p_0, p_1, \ldots, p_n\}$ spans S_n. Moreover, for each i, $p_i \notin \langle p_0, \ldots, p_{i-1} \rangle = S_{i-1}$. Thus by Lemma 2, $\{p_0, p_1, \ldots, p_n\}$ is independent, so it is a basis. □

Theorem 5. *Any finite spanning set of a projective space S contains a basis of S.*

PROOF. Let $H = \{h_0, h_1, \ldots, h_m\}$ be a finite spanning set of S. We shall construct a basis of S by deleting from H those points that depend on preceding points. If h_i depends on $h_0, \ldots, h_{i-1}$, then $\langle h_0, \ldots, h_i \rangle = \langle h_0, \ldots, h_{i-1} \rangle$, so deleting such points does not affect the span. If $h_j, j > i$, depends on $\{h_0, \ldots, h_{j-1}\}$, then h_j also depends on $\{h_0, \ldots, h_{i-1}, h_{i+1}, h_{j-1}\}$, so deletions do not affect deletions down the line. Thus we eventually arrive at a subset H' of H in which no point depends on previous points. Thus H' is independent by Lemma 2. Since the span was not affected, H' is also a spanning set of S. Thus H' is a basis of S. □

At this point we could prove that every projective space has a basis. However, the proof would involve the use of an equivalent of the axiom of choice. Rather than do so, we shall consider only certain kinds of projective spaces.

Definition 4. A projective space S is called *finite-dimensional* if and only if S has a finite spanning set.

Henceforth, we shall restrict our attention to finite-dimensional projective spaces.

By Theorem 5, a finite-dimensional projective space has a basis with a finite number of elements. It is evident that if projective space S has basis $B = \{b_0, b_1, \ldots, b_n\}$, then S is an S_n; this follows inductively from the corollary to Lemma 1. Thus every finite-dimensional projective space is an S_n for some natural number n.

If we mimic the proof of Theorem 2, we can prove the following generalization.

Theorem 6. *Every line in an S_n meets every S_{n-1} in the S_n.*

An S_{n-1} lying in an S_n is called a *hyperplane* of the S_n. Thus Theorem 6 says that in any projective space, every line meets every hyperplane.

Now we can prove two key results.

Theorem 7. *Any set of $n + 2$ points in an S_n is a dependent set.*

PROOF. Let $H = \{h_0, h_1, \ldots, h_n, h_{n+1}\}$ be a set of $n + 2$ points in an S_n, and suppose that H is independent. Then $\langle h_0, h_1, \ldots, h_{n-1}\rangle$ is a hyperplane, and h_n and h_{n+1} do not lie in it. But the line $h_n h_{n+1}$ meets the hyperplane, by Theorem 6, so that by applying Lemma 1, we have $h_{n+1} \in \langle h_0, h_1, \ldots, h_n\rangle$. Thus H is dependent after all. □

Theorem 8. *Any two bases of the same projective space have the same number of points.*

PROOF. Let $\{a_0, a_1, \ldots, a_n\}$ and $\{b_0, b_1, \ldots, b_m\}$ be bases of the same projective space S. If $n < m$, then $S = \langle a_0, \ldots, a_n\rangle$ is an S_n, and $\{b_0, \ldots, b_m\}$ is a dependent set by Theorem 7 (and the fact that a set with a dependent subset is a dependent set). Thus $m \leqslant n$. Similarly, $n \leqslant m$. Thus $m = n$, and the theorem is proved. □

Corollary. *Every basis of an S_n has $n + 1$ points.*

The number of points in a basis of a projective space S is called the *rank* of S, and the *dimension* of S is one less than the rank. Thus an S_n has dimension n. In particular, a point S_0 has dimension 0, and the empty space $\emptyset$ has rank 0 and dimension -1. Thus we may write, by convention $\emptyset = S_{-1}$.

We close this section with one more useful theorem.

Theorem 9. *If H is an independent subset of a projective space S, and B is any basis of S, then there is a subset C of B such that $H \cup C$ is a basis of S.*

PROOF. Let $B = \{b_0, b_1, \ldots, b_n\}$ be a basis of S. If $\langle H\rangle = S$, let C be the empty set. If $\langle H\rangle \neq S$, then some point of B is not in $\langle H\rangle$; let c_1 be the first point in the ordering of B that is not in $\langle H\rangle$, and let $H_1 = H \cup \{c_1\}$. If $\langle H_1\rangle = S$, let $C = \{c_1\}$. If not, let c_2 be the first point of B not in $\langle H_1\rangle$. Note that c_2 must follow c_1 in the ordering of B. Then let $H_2 = H_1 \cup \{c_2\}$; if $\langle H_2\rangle = S$, let $C = \{c_1, c_2\}$, and if not, continue. After $k \leqslant n$ steps, we arrive at the condition that no point of B fails to be in $\langle H_k\rangle$. At that stage, we have $\langle H_k\rangle = S$. Note that each H_i is independent. Thus if $C = \{c_1, \ldots, c_k\}$, $H \cup C$ is a basis of S. □

Exercises 6.1

1. Prove the following properties of spans.

(a) If $H \subseteq K$, then $\langle H \rangle \subseteq \langle K \rangle$.
(b) $\langle H \rangle \cup \langle K \rangle \subseteq \langle H \cup K \rangle$.
(c) $\langle H \cap K \rangle \subseteq \langle H \rangle \cap \langle K \rangle$.
(d) If $H \subseteq \langle K \rangle$, then $\langle H \rangle \subseteq \langle K \rangle$.
(e) If $H \subseteq \langle K \rangle$, then $\langle H \cup K \rangle = \langle K \rangle$.
(f) $\langle \langle H \rangle \rangle = \langle H \rangle$.

2. Prove Theorem 3.

3. Prove that if a subset of H is dependent, then H is dependent.

4. Prove that every subset of an independent set is independent.

5. Let A and B be subspaces of a projective space S. Let $C = \{x \in S \mid \text{for some } p \in A \text{ and some } q \in B, x \text{ is on } pq\}$. Prove that C is a subspace of S.

6. Prove that any independent set can be extended to a basis.

7. Prove the Corollary to Theorem 8.

8. Prove Theorem 6.

9. Prove that a subspace of a projective space is a projective space.

Section 6.2. Desargues's Theorem and Algebraic Examples

Recall from Chapter 3 that a Desarguesian plane is a projective plane in which any central couple of triangles is axial. The notion of a couple is still valid in a projective space of any dimension (except dimension < 2, where triangles do not exist), so we shall call a projective space S *Desarguesian* in case every central couple is axial.

We have seen examples of both Desarguesian and non-Desarguesian planes. Surprisingly enough, there are no non-Desarguesian spaces of dimension > 2.

Theorem 1. *If S is a projective space of dimension $n \neq 2$, then S is Desarguesian.*

PROOF. First, we note that if $n = -1$, 0, or 1, then S is Desarguesian by default, for no couples exist (so none can be central without being axial). Thus we need only consider $n \geqslant 3$. Let triangles abc and $a'b'c'$ form a central couple with center p. There are two cases to consider: either abc and $a'b'c'$ lie in the same plane or they don't.

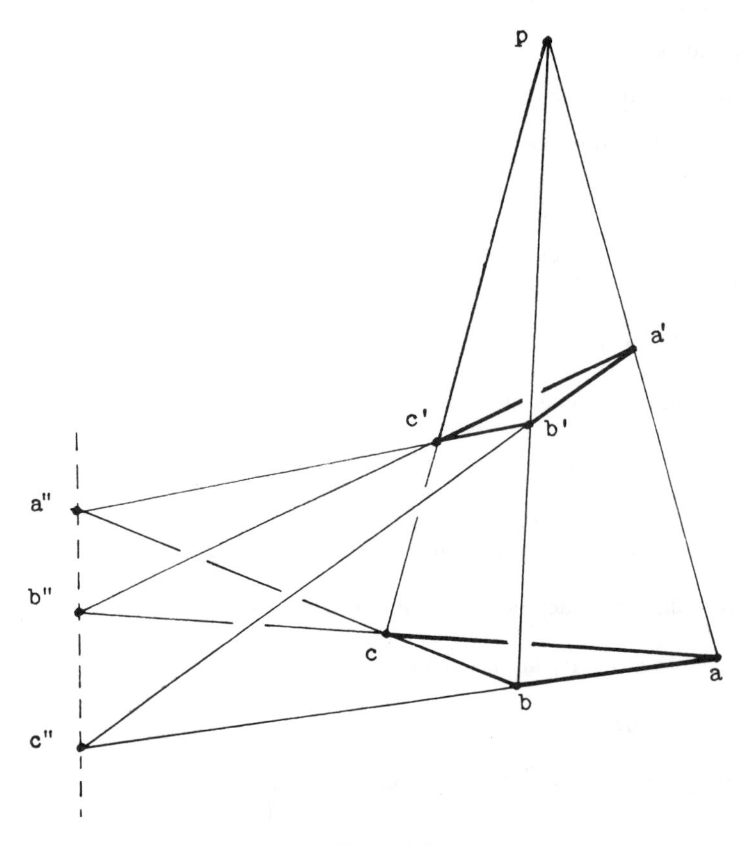

Figure 6.6

Case 1. If abc and $a'b'c'$ do not lie in the same plane (Figure 6.6), then points $a'' = bc \cap b'c'$, $b'' = ac \cap a'c'$, and $c'' = ab \cap a'b'$ all exist by S3. Moreover, because a plane is a subspace, each of a'', b'' and c'' lies in the plane of abc and in the plane of $a'b'c'$. Since two distinct planes that meet at all intersect in precisely a line, points a'', b'', and c'' are collinear, and the couple is axial.

Case 2. If abc and $a'b'c'$ do lie in the same plane, let d be a point not in that plane. Let d' be a third point on line pd (see Figure 6.7), and let $a_1 = da \cap d'a', b_1 = db \cap d'b'$, and $c_1 = dc \cap d'c'$. These points exist by S3. Then triangles abc and $a_1b_1c_1$ form a noncoplanar central couple with center d, so by Case 1, they also form an axial couple. Moreover, the axis $a''b''c''$ is the line common to the planes of abc and $a_1b_1c_1$. Similarly, triangles $a_1b_1c_1$ and $a'b'c'$ also form an axial couple with axis $a''b''c''$. Thus triangles abc and $a'b'c'$ are axial with axis $a''b''c''$ □

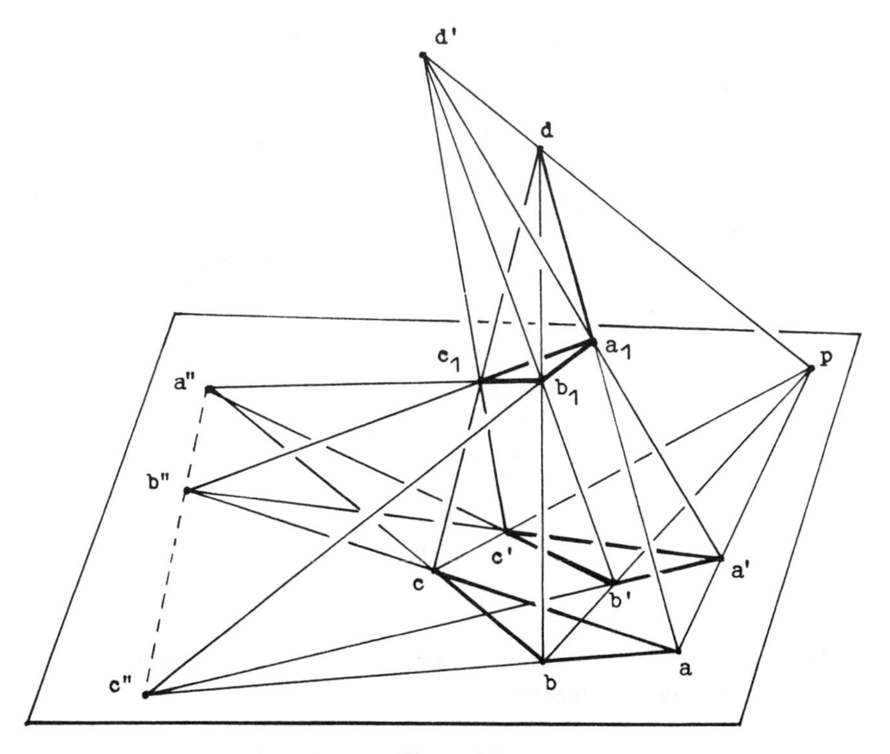

Figure 6.7

Notice that in Case 2 we are able to prove that the coplanar central couple is axial only by getting out of the plane. We therefore have the following corollary to Theorem 1.

Corollary. *A projective plane is Desarguesian if it is a subspace of another projective space.*

It will become apparent that the converse of the corollary is also true, once we develop some algebraic models.

Let D be a division ring, and let D^{n+1} be the (left) vector space of tuples over D. Let $\mathcal{P}_n$ be the set of one-dimensional (linear) subspaces of D^{n+1}; we shall denote an element of $\mathcal{P}_n$ by any nonzero vector $x = (x_0, x_1, \ldots, x_n)$ lying in it. (Thus two vectors in the same linear subspace both represent it.) Let $\mathcal{L}_n$ be the set of two-dimensional subspaces of D^{n+1}, and let $\mathcal{I}$ be set containment. The incidence structure

$$S_n(D) = (\mathcal{P}_n, \mathcal{L}_n, \mathcal{I})$$

is called the *space over D of dimension n*.

Theorem 2. *$S_n(D)$ is a projective space of dimension n.*

PROOF. Two points in $S_n(D)$ determine a line, because two linear subspaces of D^{n+1} are contained in a unique two-dimensional subspace of D^{n+1}. A line contains at least three points, because if the two-dimensional subspace L contains the two vectors x and y, then it also contains the vector $x + y$. Finally, suppose lines ab and cd meet at p. Then the two-dimensional subspace containing vectors a and b meets the two-dimensional subspace containing c and d in a one-dimensional subspace of D^{n+1} containing p. Then vectors a, c and p span a three-dimensional subspace of D^{n+1} containing b and d. In this three-dimensional subspace, the two two-dimensional subspaces ac and bd must meet, so lines ac and bd meet in $S_n(D)$. Finally, to show that $S_n(D)$ has dimension n, note that the $n + 1$ points

$$\begin{aligned} e_0 &= (1, 0, \ldots, 0), \\ e_1 &= (0, 1, 0, \ldots, 0), \\ e_2 &= (0, 0, 1, 0, \ldots, 0), \\ &\vdots \\ e_n &= (0, \ldots, 0, 1) \end{aligned}$$

form a basis of $S_n(D)$. □

Because of Theorem 1, we could imitate the development of Chapter 3 and prove the following theorem.

Theorem 3. *A projective space of dimension $n \geqslant 3$ is isomorphic to $S_n(D)$ for some division ring D.*

In Chapter 3, we proved the converse of Desargues's theorem by noticing that the converse was the dual. In a projective space of dimension $\geqslant 3$, points and lines are no longer duals of each other. But there is still a principle of duality; it is points and hyperplanes that are duals of each other. This forces a duality of subspaces in which an S_k in S_n is dual to an S_{n-k-1}.

For example, in S_3 points and planes are duals of each other, and a line is self-dual.

Algebraically, a point of $S_n(D)$ is an $(n + 1)$-tuple. Dually, a hyperplane of $S_n(D)$ is the set of points satisfying a single equation. An S_k in $S_n(D)$ is the span of $k + 1$ independent points, and its dual S_{n-k-1} is the intersection of $k + 1$ hyperplanes, or equivalently, the solution of a system of $k + 1$ simultaneous equations.

Every hyperplane of $S_n(D)$ is isomorphic to $S_{n-1}(D)$; more generally, every k-dimensional subspace of $S_n(D)$ is isomorphic to $S_k(D)$.

If E is a sub-division-ring of D, then $S_n(E)$ is a *subgeometry* of $S_n(D)$, consisting of the points of $S_n(D)$ whose coordinates can be written as elements of E.

If F is a finite field having q elements, then $S_n(F)$ is a finite projective space having $(q^{n+1}-1)/(q-1)$ points; on any line of $S_n(F)$ are $q+1$ points. A projective space with $q+1$ points per line is said to have *order q*.

If H is a hyperplane of projective space S, the set $S \setminus H$ is called a *principal restriction* of S and is an *affine space*. The affine space $S \setminus H$ bears the same relation to S as the Euclidean plane bears to the real projective plane, except that in $S \setminus H$ we have families of parallel subspaces of many dimensions.

Thus we see that all the theory of Desarguesian planes can be extended to projective spaces, where it is generalized and enriched by the addition of subspaces of various dimensions. Further developments may be found in Springer [15, Chapter 10], or in Veblen and Young [18].

Exercises 6.2

1. Prove the converse of the Corollary to Theorem 1.

2. In Case 2 in the proof of Theorem 1, explain why lines ab, $a'b'$, and a_1b_1 all pass through c''.

3. List all the points and planes in $S_3(Z_2)$.

4. Describe all the lines in $S_3(Z_2)$.

5. How many lines has a space of dimension n and order q?

6. In a space of dimension n and order q, prove

(a) there are $q+1$ points on a line,
(b) there are q^2+q+1 lines in a plane,
(c) there are $(q^n-1)/(q-1)$ lines through a point.

7. Prove that in S_n, an S_k and an S_l meet in at least an S_{k+l-n} if $k+l \geqslant n$.

8. In S_3, what is the dual configuration of

(a) a triangle,
(b) a tetrahedron?

9. In S_n, an $(n+2)$-point is a set of $n+2$ points, no $n+1$ of which lie in a hyperplane. Prove that in an $(n+2)$-point in S_n, no k points lie in an S_{k-2}.

10. In an S_n of order q, count the number of S_k's.

11. Prove that in S_3, two planes must intersect in a line.

Section 6.3 Homomorphisms

In this section, we shall introduce the notion of a homomorphism between projective spaces. We will restrict our attention to homomorphisms between spaces of the same dimension, giving some hints toward generalization in the exercises.

Definition 1. A function $f: S \to S'$ between projective spaces is a *homomorphism* if and only if given collinear points $a, b, c \ldots$ in S, points $f(a), f(b), f(c), \ldots$ in S' are also collinear.

Thus a homomorphism is any collinearity-preserving mapping. If p is a point in the domain of a homomorphism f, we shall denote the image $f(p)$ by p'.

We shall find it appropriate to consider only certain kinds of homomorphisms in this section, for a more general treatment is beyond our scope. These will be homomorphisms between spaces of the same dimension.

Definition 2. A homomorphism $f: S_n \to S_n'$ between spaces of the same dimension n is *nonsingular* if and only if the image of f contains an $(n+2)$-point of S_n'. (See Exercise 6.2.9).

Hereafter, unless specifically stated otherwise, all homomorphisms will be nonsingular, and between spaces of dimension n.

If $a_0a_1 \cdots a_{n+1}$ is an $(n+2)$-point in the domain of homomorphism f, and its image $a_0'a_1' \cdots a_{n+1}'$ is also an $(n+2)$-point, we say f *preserves* the $(n+2)$-point $a_0a_1 \cdots a_{n+1}$. It is evident that every (nonsingular) homomorphism preserves some $(n+2)$-point.

To give the flavor of the theorems that will follow, we first consider what a homomorphism does to certain points associated with an $(n+2)$-point Let $a_0a_1 \cdots a_{n+1}$ be an $(n+2)$-point. A *diagonal point* of $a_0a_1 \cdots a_{n+1}$ is a point collinear with two of the points $a_0, a_1, \ldots, a_{n+1}$ and also in the hyperplane spanned by the remaining $n-1$ points. We let H_{ij} be the hyperplane spanned by $\{a_0, a_1, \ldots, a_{n+1}\}\backslash\{a_i, a_j\}$, and we let d_{ij} denote the diagonal point $a_ia_j \cap H_{ij}$.

Now suppose f is a homomorphism that preserves $a_0a_1 \cdots a_{n+1}$. If a_i and a_j are any two of the points, and p is any point on line a_ia_j, then its image p' is on line $a_i'a_j'$. This is because f preserves collinearity. Next, let a_i, a_j, a_k be three of the points, and let p be any point in their span $\langle a_i, a_j, a_k \rangle$. Let $p_i = a_ip \cap a_ja_k$. Then p_i' is on $a_j'a_k'$, and p' is on $a_i'p_i'$; thus p' is in the span $\langle a_i', a_j', a_k' \rangle$. Thus in general, if $p \in \langle a_{i_1}, a_{i_2}, \ldots, a_{i_k} \rangle$, then $p' \in \langle a_{i_1}', a_{i_2}, \ldots, a_{i_k}' \rangle$.

It now follows that if d_{ij} is a diagonal point of $a_0a_1 \cdots a_{n+1}$, then d_{ij}' is a diagonal point of $a_0'a_1' \cdots a_{n+1}'$. Thus the homomorphism f that preserves

$a_0a_1 \cdots a_{n+1}$ carries the diagonal points of $a_0a_1 \cdots a_{n+1}$ onto the diagonal points of $a_0'a_1' \cdots a_{n+1}'$.

Our first theorem shows that a homomorphism preserves many $(n+2)$-points.

Theorem 1. *If $f: S_n \to S_n'$ is a homomorphism and $x \in S_n$, then there exist $n+1$ points $a_1, a_2, \ldots, a_{n+1} \in S_n$ such that $xa_1a_2 \cdots a_{n+1}$ is an $(n+2)$-point preserved by f.*

PROOF. We shall prove this theorem by induction on the dimension n of the spaces involved. First of all, if $f: S_1 \to S_1'$ is a homomorphism, then f preserves a three-point, which means that f carries some three points in S_1 into three distinct points of S_1'. Suppose a_0, a_1, a_2 are three points with distinct images. If x is any point in S_1, then x' is distinct from two of a_0', a_1', a_2', say the last two. Then xa_1a_2 is a three-point preserved by f. Thus the theorem holds for $n = 1$.

Now suppose the theorem holds for $n-1$; we prove it holds for n. Suppose $f: S_n \to S_n'$ preserves the $(n+2)$-point $a_0a_1 \cdots a_{n+1}$, and let $x \in S_n$. Any n of the points $a_0', a_1', \ldots, a_{n+1}'$ span a hyperplane of S_n'; there must be one of them that does not contain x'. Suppose $\langle a_2', a_3', \ldots, a_{n+1}' \rangle$ is a hyperplane not containing x'; then $x \notin \langle a_2, a_3, \ldots, a_{n+1} \rangle$. Now $d_{01}a_2a_3 \cdots a_{n+1}$ is an $(n+1)$-point in the $(n-1)$-dimensional subspace $\langle a_2, \ldots, a_{n+1} \rangle$ that is preserved by f. Now x' must be distinct from either a_0' or a_1'; say $x' \neq a_0'$. Let line a_0x meet the hyperplane $\langle a_2, \ldots, a_{n+1} \rangle$ at point y. Then, by the inductive hypothesis, there are points $b_1, \ldots, b_n$ in $\langle a_2, \ldots, a_{n+1} \rangle$ such that $yb_1 \cdots b_n$ is an $(n+1)$-point preserved by f. Hence $xa_0b_1 \cdots b_n$ is an $(n+2)$-point preserved by f, and the proof is complete. □

From Theorem 1 follows a more general theorem, whose proof we omit.

Theorem 2. *If $f: S_n \to S_n'$ is a homomorphism and $x_0, \ldots, x_k \in S_n$ are such that $x_0', \ldots, x_k'$ are independent, then there exist $n-k+1$ points $a_0, a_1, \ldots, a_{n-k} \in S_n$ such that $x_0x_1 \cdots x_ka_0a_1 \cdots a_{n-k}$ is an $(n+2)$-point preserved by f.*

That is, any set of points whose images under f are independent may be extended to an $(n+2)$-point preserved by f.

If $f: S_n \to S_n'$ is a homomorphism, we shall say two points $x, y \in S_n$ are *congruent modulo* f in case $x' = y'$. It is easily verified that congruence modulo f is an equivalence relation; the equivalence class of point x is denoted by $[x]_f$, and is called the *f-class* of x. If f is surjective, the f-classes in S_n form a projective geometry isomorphic to S_n'; the first step in demonstrating that fact is the following theorem.

Theorem 3. *If $f: S_n \to S_n'$ is a homomorphism, then any S_k in S_n meets at least $k+2$ f-classes.*

PROOF. We give the argument for $k = 1$ and leave the rest as an exercise. Let S_1 be a line in S_n, and let p be a point on S_1. Let $pa_1 \cdots a_{n+1}$ be an $(n+2)$-point preserved by f. If any a_i is on S_1, then S_1 contains p, a_i, and a diagonal point of the $(n+2)$-point, all of whose images are distinct. Thus S_1 meets three f-classes. If no a_i is on S_1, let each hyperplane $H_i = \langle\{a_1, \ldots, a_{n+1})\}\backslash\{a_i\}\rangle$ meet S_1 at point q_i. Since at most $n-1$ of the H_i's contain S_1, two of the q_i's must be distinct (else some a_j is on S_1). They and p also have distinct images, so again S_1 meets three f-classes. □

To tell us a little more about f-classes, we have the following result.

Theorem 4. *If a homomorphism $f: S_n \to S_n'$ preserves the $(n+2)$-point $pa_1 \cdots a_{n+1}$, and q and r are two points on a_1a_2 in the same f-class, then there exist distinct points x and y that are in the same f-class as p.*

PROOF. One of the hyperplanes $H_1 = \langle a_1, a_3, \ldots, a_{n+1}\rangle$ and $H_2 = \langle a_2, a_3, \ldots, a_{n+1}\rangle$ contains neither q nor r, for if $q \in H_1$, then $q' \in a_1'a_2' \cap H_1' = a_1' \Rightarrow r' = a_1' \Rightarrow q, r \notin H_2$. Suppose H_1 contains neither q nor r, and let $q_1 = pq \cap H_1$, $r_1 = pr \cap H_1$. Also let $H_3 = \langle p, a_3, \ldots, a_{n+1}\rangle$, and let $x = q_1r \cap H_3$, $y = qr_1 \cap H_3$. Then $q' = r' \Rightarrow q_1' = p'q' \cap H_1' = p'r' \cap H_1' = r_1'$. If $q' \neq a_1'$, then $x' = q_1'r' \cap H_3' = r_1'q' \cap H_3' = y'$; also $x' = q_1'r' \cap H_3' = r_1'r' \cap H_3' = p'$. Thus x and y are in the same f-class as p. If $q' = a_1'$, then H_2 contains neither q nor r; replace H_1 by H_2 and proceed as at first. □

Theorem 4 tells us that f-classes are somewhat alike. The next two theorems make that observation more precise, and apply it to a characterization of f.

Theorem 5. *If $f : S_n \to S_n'$ is a homomorphism, then the following are equivalent:*

1. *f is injective.*
2. *f preserves every $(n+2)$-point in S_n.*
3. *f preserves noncollinearity.*
4. *For some $p \in S_n$, $[p]_f = \{p\}$.*

PROOF. $1 \Rightarrow 2$: By induction on n. If $n = 1$, the images of three distinct points are three distinct points, because f is injective. Suppose a homomorphism that is injective preserves every $(n+1)$-point in S_{n-1}, and let $a_0 \cdots a_{n+1}$ be an $(n+2)$-point in S_n. If $a_0', \ldots, a_{n+1}'$ do not form an $(n+2)$-point in S_n', then some $n+1$ of these points lie in a hyperplane. Suppose $a_0', \ldots, a_n'$ lie in a hyperplane H' in S_n'. Now $a_0, \ldots, a_{n-1}$ span a hyperplane H in S_n, by Exercise 6.2.9. Let $p = a_na_{n+1} \cap H$; then $a_0a_1 \cdots a_{n-1}p$ is an $(n+1)$-point in H, and so is preserved by f, by the inductive hypothesis. Thus $a_0' \cdots a_{n-1}'p'$ is an $(n+1)$-point in H'. In particular, $\langle a_0', \ldots, a_{n-1}'\rangle = H'$. Let $a_0 \cdots a_{n-1}b_nb_{n+1}$ be an $(n+2)$-point in S_n preserved by f, which exists by Theorem 2. Let $x = b_nb_{n+1} \cap \langle a_0, \ldots, a_{n-1}\rangle$ and $y = b_nb_{n+1} \cap \langle a_1, \ldots, a_n\rangle$. Then $x \neq y$, but $x' = b_n'b_{n+1}' \cap \langle a_0', \ldots, a_{n-1}'\rangle = b_n'b_{n+1}' \cap H'$

$= b_n'b_{n+1}' \cap \langle a_1' \cdots a_n' \rangle = y'$, contradicting the fact that f is injective. Hence f preserves every $(n+2)$-point.

$2 \Rightarrow 3$, $3 \Rightarrow 1$, and $1 \Rightarrow 4$ are left as exercises.

$4 \Rightarrow 1$: Let $p \in S_n$ such that $[p]_f = \{p\}$. Suppose x and y are distinct points in S_n such that $x' = y'$. Let $pa \cdots a_{n+1}$ be an $(n+2)$-point preserved by f, and let $H_i = \langle \{a_1, \ldots, a_{n+1}\} \backslash \{a_i\} \rangle$. Then there are distinct i, j such that neither H_i nor H_j contains either x or y (otherwise $x = y$).

Let

$$q = a_i a_j \cap \langle \{x, a_1, \ldots, a_{n+1}\} \backslash \{a_i, a_j\} \rangle,$$

$$r = a_i a_j \cap \langle \{y, a_1, \ldots, a_{n+1}\} \backslash \{a_i, a_j\} \rangle.$$

Then

$$\begin{aligned} q' &= a_i' a_j' \cap \langle \{x', a_1', \ldots, a_{n+1}'\} \backslash \{a_1, a_j'\} \rangle \\ &= a_i' a_j' \cap \langle \{y', a_1', \ldots, a_{n+1}\} \backslash \{a_i', a_j'\} \rangle \\ &= r'. \end{aligned}$$

Hence by Theorem 4, $[p]_f \neq \{p\}$; this contradiction forces f to be injective. □

Theorem 6. *If $f: S_n \to S_n'$ $(n > 1)$ is a homomorphism and f is not injective, then for any $p \in S_n$, $[p]_f$ contains an $(n+2)$-point of S_n.*

PROOF. By induction on n. If $n = 2$, let $p \in S_2$. By part 4 of Theorem 5, there exists $b_1 \in [p]_f$, $b_1 \neq p$. Let $pa_1a_2a_3$ be a four-point preserved by f. We may suppose without loss of generality that b_1 is not on pa_1, pa_2, or a_1a_2. Let $b_2 = a_1b_1 \cap pa_2$ and $b_3 = a_2b_1 \cap pa_1$. Then $pb_1b_2b_3$ is a four-point and $b_1, b_2, b_3 \in [p]_f$.

Assume the theorem is true for $n-1$. Let $f: S_n \to S_n'$ be a noninjective homomorphism, let $p \in S_n$, and let H be a hyperplane of S_n that contains p. Then by the inductive hypothesis, there is an $(n+1)$-point $pb_1 \cdots b_n$ in H with $b_1, \ldots, b_n \in [p]_f$. Then $\{p, b_1, \ldots, b_{n-1}\}$ spans H. Let $pa_1 \cdots a_{n+1}$ be an $(n+2)$-point in S_n that is preserved by f. Then some two of the points $a_1, \ldots, a_{n+1}$ are not in H, say a_i and a_j. Let $c_n = \langle a_i, b_1, \ldots, b_{n-1} \rangle \cap pa_j$ and let $c_{n+1} = \langle a_j, b_1, \ldots, b_{n-1} \rangle \cap pa_i$. Then $pb_1 \cdots b_{n-1}c_nc_{n+1}$ is an $(n+2)$-point. Also, $c_n' = \langle a_i', b_1', \ldots, b_{n-1}' \rangle \cap p'a_i' = a_i'p' \cap p'a_j' = p'$, and similarly $c_{n+1}' = p'$. Thus the theorem is proved. □

We now see that a homomorphism either preserves all $(n+2)$-points or else carries "lots" of $(n+2)$-points to single points. In order to see how this latter event is possible, we shall look at some algebraic examples.

First, let us recall (Theorem 6.2.3) that for $n > 2$, S_n is isomorphic to $S_n(D)$ for some division ring D. That is, given S_n and any $(n+2)$-point $a_0a_1 \cdots a_{n+1}$, there is an isomorphism $f: S_n \to S_n(D)$ such that $f(a_i) = (0, \ldots, 0, 1, 0 \cdots 0)$ (1 in the ith position) for $i = 0, \ldots, n$, and $f(a_{n+1}) = (1, 1, \ldots, 1)$. We call $a_0 \cdots a_{n+1}$ a *coordinate basis* of S_n.

And just what is an isomorphism? It is precisely a bijective homomorphism. In particular, a collineation on S_n is just a bijective homomorphism $f: S_n \to S_n$.

We now consider homomorphisms $f: S_n(D) \to S_n(D')$ between projective spaces over division rings D and D'. The coordinate basis of $S_n(D)$ will be taken as $e_0 = (1, 0, \ldots, 0)$, $e_1 = (0, 1, 0, \ldots, 0), \ldots, e_n = (0, \ldots, 0, 1)$, $e_{n+1} = (1, 1, \ldots, 1)$. A homomorphism $f: S_n(D) \to S_n(D')$ will be called *basic* in case the coordinate basis of $S_n(D)$ is carried (in order) onto the coordinate basis of $S_n(D')$.

Let D and D' be division rings, and let ∞ be an element not in D'. We extend addition and multiplication from D' to $D' \cup \{\infty\}$ by setting, for $a \in D'$, $a \neq 0$,

$$a + \infty = \infty, \qquad \infty + \infty = \infty,$$
$$a\infty = \infty, \qquad \infty\,\infty = \infty.$$

A function $\phi: D \to D' \cup \{\infty\}$ is a *place* if and only if

1. $\phi(a+b) = \phi(a) + \phi(b) \quad [a + b \neq 0 \text{ if } \phi(a) = \infty]$,
2. $\phi(ab) = \phi(a)\phi(b)$,
3. if $\phi(a) = \infty$, then $\phi(a^{-1}) = 0$.

We denote

$$A_\phi = \{x \in D \mid \phi(x) \neq \infty\},$$
$$J_\phi = \{x \in D \mid \phi(x) = 0\}.$$

The following theorems can now be proved; their proofs are heavily dependent on algebraic properties of A_ϕ and J_ϕ, however, and will not be given. For proofs, see André [2], Dress [8], Garner [10], Klingenberg [13], and Radó [14].

Theorem 7. *If $\phi : D \to D' \cup \{\infty\}$ is a place and $x_0, x_1, x_2 \cdots \in D$ are not all zero, then there exists $b \in D$ such that $bx_0, bx_1, bx_2 \ldots$ are all in A_ϕ but not all in J_ϕ.*

The effect of Theorem 7 is that a point $(x_0, x_1, \ldots, x_n) \in S_n(D)$ can be written (only proportionality counts) so that all its entries are in A_ϕ [so $\phi(x_i) \in D'$] but not all in J_ϕ [so $\phi(x_i)$ is not zero for all i]. Thus the place ϕ defines a function

$$f_\phi : S_n(D) \to S_n(D')$$

by the formula

$$f_\phi(x_0, x_1, \ldots, x_n) = (\phi(x_0), \phi(x_1), \ldots, \phi(x_n)).$$

Theorem 8. *If $\phi: D \to D' \cup \{\infty\}$ is a place, then the function $f_\phi: S_n(D) \to S_n(D')$ defined above is a basic homomorphism.*

Theorem 9. *If $f: S_n(D) \to S_n(D')$ is a basic homomorphism, then there is a unique place $\phi_f: D \to D' \cup \{\infty\}$ such that $f_{\phi_f} = f$.*

Now for the promised examples.

Example 1. Let Q be the field of rational numbers, and let p be a prime number. Let $A_p = \{r/s \in Q \mid p \text{ is not a factor of } s\}$. Let $\phi_p : Q \to Z_p \cup \{\infty\}$, with Z_p the field of integers modulo p, be defined by

$$\phi_p\left(\frac{r}{s}\right) = \begin{cases} r \pmod p & \text{if } \frac{r}{s} \in A_p, \\ \infty & \text{if } \frac{r}{s} \notin A_p. \end{cases}$$

Then $f_{\phi_p} : S_n(Q) \to S_n(Z_p)$ is a homomorphism from an infinite projective geometry into a finite projective geometry.

Example 2. It is rather surprising that the composition of homomorphisms may be singular. Let R be the field of real numbers, and let $K = R(t)$ be the field of rational functions in t over R. Let $g : S_n(R) \to S_n(K)$ be defined by $g(x_0, x_1, \ldots, x_n) = (x_0, x_1 t, \ldots, x_n t)$. Then g is a homomorphism. In K, let

$$A = \left\{ \frac{p(t)}{q(t)} \in K \,\middle|\, \text{degree of } p \leqslant \text{degree of } q \right\}.$$

Let $\phi : K \to R$ be defined by

$$\phi\left(\frac{p(t)}{q(t)}\right) = \begin{cases} \lim\limits_{t\to\infty} \dfrac{p(t)}{q(t)} & \text{if } \dfrac{p(t)}{q(t)} \in A, \\ \infty & \text{if } \dfrac{p(t)}{q(t)} \notin A. \end{cases}$$

Then ϕ is a place, so that $f_\phi : S_n(K) \to S_n(R)$ is a homomorphism. But now

$$f_\phi \cdot g(x_0, x_1, \ldots, x_n) = f(x_0, x, t, \ldots, x_n t)$$

$$= \begin{cases} f_\phi(1, 0, \ldots, 0) = (1, 0, \ldots, 0) & \text{if } x_1 = \cdots = x_n = 0 \\ f_\phi\left(\dfrac{x_0}{t}, x_1, \ldots, x_n\right) = (0, x_1, \ldots, x_n) & \text{otherwise.} \end{cases}$$

Hence the image of $f_\phi \cdot g : S_n(R) \to S_n(R)$ consists of the point $(1, 0, \ldots, 0)$ and the hyperplane $x_1 = 0$, which contains no $(n+2)$-point. Thus $f_\phi \cdot g$ is singular.

Exercises 6.3

1. Explain why every nonsingular homomorphism preserves an (n + 2)-point.

2. Prove Theorem 2.

3. Prove Theorem 3 in the general case.

4. Complete the proof of Theorem 5.

5. Let $\phi : D \to D' \cup \{\infty\}$ be a place. Prove that if $x \in D \backslash A_\phi$, then $x^{-1} \in J_\phi$.

6. Prove that if ϕ: $D \to D' \cup \{\infty\}$ is a place, $x \in A_\phi$, and $y \in J_\phi$, then $xy \in J_\phi$ and $yx \in J_\phi$.

7. If ϕ: $D \to D' \cup \{\infty\}$ is a place and $x \in A_\phi \backslash J_\phi$, prove that $x^{-1} \in A_\phi$.

8. Let ϕ_2: $Q \to Z_2 \cup \{\infty\}$ be a place as in Example 1, with $p = 2$. Find the images under f_{ϕ_2} of the following points in $S_3(Q)$:

(a) $(1, 2, \frac{1}{3}, 0)$, **(b)** $(0, 1, 1, \frac{1}{2})$,
(c) $(\frac{1}{2}, \frac{1}{3}, \frac{1}{4}, \frac{1}{5})$, **(d)** $(-1, 1, -\frac{1}{2}, \frac{1}{2})$.

9. Show that congruence modulo f is an equivalence relation on S_n when f: $S_n \to S_n'$ is a homomorphism.

10. The singular homomorphism $f_\phi \cdot g$ of Example 2 is called a *projection* from the point $(1, 0, \ldots, 0)$ onto the hyperplane $x_1 = 0$. Describe a more general projection from the S_k $x_{k+1} = 0, \ldots, x_n = 0$ onto the S_{n-k-1} $x_1 = 0, \ldots, x_k = 0$, and give it as a composition of homomorphisms.

11. Generalize the theory of collineations on a projective plane (Section 2.3) to collineations on S_n. (See, for example, Verdina [19].)

12. Two homomorphisms $f, g : S_n \to S_n'$ are *equivalent* if and only there exist projective collineations $h : S_n \to S_n$ and $h' : S_n' \to S_n'$ such that $h' \cdot f = g \cdot h$. Prove that this equivalence is an equivalence relation.

13. Prove that if homomorphisms f, g : $S_n(D) \to S_n(D')$ are basic and equivalent, then $f = g$.

14. Provc that a homomorphism f : $S_n(D) \to S_n(D')$ is equivalent to a unique basic homomorphism f : $S_n(D) \to S_n(D')$.

15. Prove that if homomorphisms f and g are equivalent, then f is injective (surjective) if and only if g is injective (surjective).

16. Prove that for each homomorphism f : $S_n(D) \to S_n(D')$ there is a unique place ϕ_f : $D \to D' \cup \{\infty\}$ such that f is equivalent to f_{ϕ_f}.

17. Prove that the homomorphism f is injective (surjective) if and only if ϕ_f in Exercise 16 is injective (surjective).

18. Prove that f : $S_n(D) \to S_n(D)$ is a collineation if and only if ϕ_f : $D \to D \cup \{\infty\}$ is an automorphism of D. (It can also be shown that f is a projective collineation if and only if ϕ_f is an inner automorphism of D; see Stevenson [16, p. 246].)

19. Let $f: S \to S'$ be a (not necessarily nonsingular) homomorphism. We say f is *k-singular* if $\dim(S) - \dim(\langle f(S) \rangle) = k$. Show that a projection from a point to a hyperplane is 0-singular, and the projection from an S_k onto an S_{n-k-1} is $(k-1)$-singular.

20. The *quotient space* of an S_{k-1} in S_n is the space S_n / S_{k-1} whose points are the subspaces of S_n of which S_{k-1} is a hyperplane. Let $f: S_n \to S_n \backslash S_{k-1}$ be defined by

$$f(x) = \begin{cases} S_{k-1} & \text{if} \quad x \in S_{k-1}, \\ \langle S_{k-1} \cup \{x\} \rangle & \text{if} \quad x \notin S_{k-1}. \end{cases}$$

Show that f is a k-singular homomorphism.

21. If S_n is a subspace of S_m, show that the inclusion map $f: S_n \to S_m$, defined by $f(x) = x$, is a homomorphism.

Appendix

Topics from Linear and Abstract Algebra

In this appendix we mention for reference the basic ideas from linear algebra and modern abstract algebra used in this book. For a more extensive treatment, see textbooks on those subjects, such as Agnew and Knapp [1] and Herstein [11].

Matrices

A *matrix* is a rectangular array of numbers. A matrix with m rows and n columns is called an $m \times n$ matrix, and may be represented by

$$A = \begin{bmatrix} a_{11} & a_{12} & \cdots & a_{1n} \\ a_{21} & a_{22} & \cdots & a_{2n} \\ \vdots & \vdots & & \vdots \\ a_{m1} & a_{m2} & \cdots & a_{mn} \end{bmatrix}.$$

The element a_{ij} represents the entry in the ith row and jth column. We often abbreviate the above array by

$$A = (a_{ij}).$$

A *square* matrix of order n is an $n \times n$ matrix. A $1 \times n$ matrix

$$X = (x_1, \ldots, x_n)$$

is called a *row vector* of *length* n. An $n \times 1$ matrix

$$Y = \begin{bmatrix} y_1 \\ \vdots \\ y_n \end{bmatrix}$$

is called a *column vector* of *length* n. A 1×1 matrix is just a number, usually referred to as a *scalar*.

Two matrices $A = (a_{ij})$ and $B = (b_{ij})$ of the same size are *equal* if and only if $a_{ij} = b_{ij}$ for all i and j.

The *transpose* of an $m \times n$ matrix $A = (a_{ij})$ is the $n \times m$ matrix

$$A^{\mathrm{T}} = (a_{ji})$$

obtained by writing the rows of A as columns (and hence the columns as rows). Thus the transpose of a row vector is a column vector, and vice versa. Also,

$$(A^{\mathrm{T}})^{\mathrm{T}} = A.$$

A matrix is *symmetric* if and only if $A^{\mathrm{T}} = A$, or $a_{ij} = a_{ji}$ for all i and j. Clearly, a symmetric matrix must be square.

If $A = (a_{ij})$ and $B = (b_{ij})$ are matrices of the same size, the *sum* of A and B is the matrix

$$A + B = (a_{ij} + b_{ij})$$

obtained by adding the corresponding entries of A and B. If A and B are not the same size, their sum is not defined.

If c is a scalar and $A = (a_{ij})$ is a matrix of any size, the *scalar product* of c and A is the matrix

$$cA = (ca_{ij})$$

obtained by multiplying each entry of A by c. The scalar product $(-1)A$ is written as $-A$, and is called the *negative* of A. The *difference* of matrices A and B is the matrix

$$A - B = A + (-B)$$

and is defined when A and B are the same size.

If $A = (a_{ij})$ is an $m \times n$ matrix and $B = (b_{ij})$ is an $n \times p$ matrix, the *product* of A and B is the $m \times p$ matrix

$$AB = (c_{ij})$$

with

$$c_{ij} = a_{i1}b_{1j} + a_{i2}b_{2j} + \cdots + a_{in}b_{nj}.$$

That is, the entry in the ith row and jth column of AB is the sum of the products of the elements in the ith row of A with the corresponding elements in the jth column of B. If the number of columns of A is not the same as the number of rows of B, then the product AB is not defined.

The *nth order identity matrix* is the $n \times n$ matrix

$$I = \begin{bmatrix} 1 & 0 & \cdots & 0 \\ 0 & 1 & \cdots & 0 \\ \vdots & \vdots & \ddots & \vdots \\ 0 & 0 & \cdots & 1 \end{bmatrix}$$

with $a_{ij} = 1$ for each i and $a_{ij} = 0$ for $i \neq j$. It is easily verified that $AI = A$ and $IB = B$ for any $m \times n$ matrix A and any $n \times p$ matrix B. Any scalar multiple of the identity matrix is called a *scalar matrix*.

The following properties of sums, products, and scalar products are easily shown:

1. $A + (B + C) = (A + B) + C$,
2. $A + B = B + A$,
3. $(bc)A = b(cA)$,
4. $c(A + B) = cA + cB$,
5. $(AB)C = A(BC)$,
6. $A(B + C) = AB + AC$,
7. $(A + B)C = AC + BC$,
8. $c(AB) = (cA)B = A(cB)$,
9. $(A + B)^{\mathrm{T}} = A^{\mathrm{T}} + B^{\mathrm{T}}$,
10. $(cA)^{\mathrm{T}} = c(A^{\mathrm{T}})$,
11. $(AB)^{\mathrm{T}} = B^{\mathrm{T}}A^{\mathrm{T}}$.

Determinants

Associated with each square matrix A is a number denoted by $|A|$ or det (A), called the *determinant* of A. If

$$A = \begin{pmatrix} a_{11} & a_{12} & \cdots & a_{1n} \\ a_{21} & a_{22} & \cdots & a_{2n} \\ \vdots & \vdots & & \vdots \\ a_{n1} & a_{n2} & \cdots & a_{nn} \end{pmatrix},$$

we also write

$$|A| = \begin{vmatrix} a_{11} & a_{12} & \cdots & a_{1n} \\ a_{21} & a_{22} & \cdots & a_{2n} \\ \vdots & \vdots & & \vdots \\ a_{n1} & a_{n2} & \cdots & a_{nn} \end{vmatrix}.$$

The definition of determinant requires some technical terminology, which we develop now. Until further notice, all matrices mentioned are square.

Suppose $A = (a_{ij})$ is a matrix of order n. The *minor* of the element a_{ij} is the matrix M_{ij} of order $n - 1$ obtained from A by deleting the ith row and jth column. The *cofactor* of a_{ij} is the number

$$A_{ij} = (-1)^{i+j} |M_{ij}|.$$

We can now define the determinant of a 1×1 matrix $A = (a_{11})$ to be

$$|A| = a_{11}.$$

For $n > 1$, the determinant of an $n \times n$ matrix $A = (a_{ij})$ is the sum

$$|A| = a_{11}A_{11} + a_{12}A_{12} + \cdots + a_{1n}A_{1n}$$

of the products of the elements in the first row of A with their respective cofactors.

Thus, for a 2×2 determinant,

$$\begin{vmatrix} a_{11} & a_{12} \\ a_{21} & a_{22} \end{vmatrix} = a_{11}a_{22} - a_{12}a_{21},$$

and for a 3×3 determinant,

$$\begin{vmatrix} a_{11} & a_{12} & a_{13} \\ a_{21} & a_{22} & a_{23} \\ a_{31} & a_{32} & a_{33} \end{vmatrix} = a_{11}\begin{vmatrix} a_{22} & a_{23} \\ a_{32} & a_{33} \end{vmatrix} - a_{12}\begin{vmatrix} a_{21} & a_{23} \\ a_{31} & a_{23} \end{vmatrix} + a_{13}\begin{vmatrix} a_{21} & a_{22} \\ a_{31} & a_{32} \end{vmatrix}$$

$$= a_{11}(a_{22}a_{33} - a_{23}a_{32}) - a_{12}(a_{21}a_{33} - a_{23}a_{31}) + a_{13}(a_{21}a_{32} - a_{22}a_{31})$$
$$= a_{11}a_{22}a_{33} - a_{11}a_{23}a_{32} - a_{12}a_{21}a_{33} + a_{12}a_{23}a_{31} + a_{13}a_{21}a_{32} - a_{13}a_{22}a_{31}.$$

It can then be shown, by expanding, that for any square matrix, A, $|A|$ can be computed as the sum of the products of the elements of any row or column with their respective cofactors. That is,

$$|A| = a_{i1}A_{i1} + a_{i2}A_{i2} + \cdots + a_{in}A_{in}$$

and

$$|A| = a_{1j}A_{1j} + a_{2j}A_{2j} + \cdots + a_{nj}A_{nj}$$

for any i and j.

As a consequence, if some row or column of A consists entirely of zeros, then $|A| = 0$.

Another consequence is that if the elements of one row or column of A are each multiplied by the constant c, then $|A|$ is multiplied by c. Schematically, we have

$$\begin{vmatrix} a_{11} & a_{12} & \cdots & a_{1n} \\ \vdots & \vdots & & \vdots \\ ca_{i1} & ca_{i2} & \cdots & ca_{in} \\ \vdots & \vdots & & \vdots \\ a_{n1} & a_{n2} & \cdots & a_{nn} \end{vmatrix} = c\begin{vmatrix} a_{11} & a_{12} & \cdots & a_{1n} \\ \vdots & \vdots & & \vdots \\ a_{i1} & a_{i2} & \cdots & a_{in} \\ \vdots & \vdots & & \vdots \\ a_{n1} & a_{n2} & \cdots & a_{nn} \end{vmatrix},$$

and similarly for columns.

A third consequence is that if one row or column of A consists of sums of two elements, then the determinant of A can be written as the sum of two determinants. Schematically, we have

$$\begin{vmatrix} a_{11} & a_{12} & \cdots & a_{1n} \\ \vdots & \vdots & & \vdots \\ b_{i1}+c_{i1} & b_{i2}+c_{i2} & \cdots & b_{in}+c_{in} \\ \vdots & \vdots & & \vdots \\ a_{n1} & a_{n2} & \cdots & a_{nn} \end{vmatrix} = \begin{vmatrix} a_{11} & a_{12} & \cdots & a_{1n} \\ \vdots & \vdots & & \vdots \\ b_{i1} & b_{i2} & \cdots & b_{in} \\ \vdots & \vdots & & \vdots \\ a_{n1} & a_{n2} & \cdots & a_{nn} \end{vmatrix} + \begin{vmatrix} a_{11} & a_{12} & \cdots & a_{1n} \\ \vdots & \vdots & & \vdots \\ c_{i1} & c_{i2} & \cdots & c_{in} \\ \vdots & \vdots & & \vdots \\ a_{n1} & a_{n2} & \cdots & a_{nn} \end{vmatrix},$$

and similarly for columns.

As yet another consequence, it can be shown that a matrix and its transpose have the same determinant, or

$$|A^{\mathrm{T}}| = |A|.$$

Using an inductive argument, it is easily shown that if two rows (or columns) of a matrix A are interchanged, then the sign of the determinant is changed. It follows from this fact that if two rows (or columns) of a matrix A are identical, then $|A| = 0$. For interchanging the identical rows changes the sign, and at the same time leaves everything unchanged.

If the elements of one row (or column) of a matrix A are multiplied by the cofactors of the corresponding elements of another row (or column), the result is zero. That is because the result is the expansion of a determinant with two identical rows (or columns).

If the elements of one row (or column, throughout) of a matrix A are replaced by the sums of those elements and a constant multiple of the corresponding elements of another row, the determinant is unchanged. Schematically, we have

$$\begin{vmatrix} a_{11} & a_{12} & \cdots & a_{1n} \\ \vdots & \vdots & & \vdots \\ a_{i1} & a_{i2} & \cdots & a_{in} \\ \vdots & \vdots & & \vdots \\ a_{j1}+ka_{i1} & a_{j2}+ka_{i2} & \cdots & a_{jn}+ka_{in} \\ \vdots & \vdots & & \vdots \\ a_{n1} & a_{n2} & \cdots & a_{nn} \end{vmatrix} = \begin{vmatrix} a_{11} & a_{12} & \cdots & a_{1n} \\ \vdots & \vdots & & \vdots \\ a_{i1} & a_{i2} & \cdots & a_{in} \\ \vdots & \vdots & & \vdots \\ a_{j1} & a_{j2} & \cdots & a_{jn} \\ \vdots & \vdots & & \vdots \\ a_{n1} & a_{n2} & \cdots & a_{nn} \end{vmatrix} + k \begin{vmatrix} a_{11} & a_{12} & \cdots & a_{1n} \\ \vdots & \vdots & & \vdots \\ a_{i1} & a_{i2} & \cdots & a_{in} \\ \vdots & \vdots & & \vdots \\ a_{i1} & a_{i2} & \cdots & a_{in} \\ \vdots & \vdots & & \vdots \\ a_{n1} & a_{n2} & \cdots & a_{nn} \end{vmatrix},$$

and the last determinant is zero.

One final fact about determinants that is not so easily seen is that the determinant of a product is the product of the determinants, or

$$|AB| = |A||B|.$$

If A is a matrix whose determinant is zero, we say A is *singular*. If $|A| \neq 0$, we say A is *nonsingular*. By the above property of determinants, it follows that the product of nonsingular matrices is nonsingular.

If A is a matrix, not necessarily square, a *subdeterminant* of A is the determinant of a (square) matrix obtained from A by deleting any number of rows and/or columns. The *rank* of A is the order of the largest nonzero subdeterminant of A. Thus an $n \times n$ matrix of rank n is nonsingular, while an $n \times n$ matrix of rank less than n must be singular.

Inverse of a Matrix

The inverse of a square matrix A is a matrix A^{-1} such that $AA^{-1} = I$ and $A^{-1}A = I$. Not all square matrices have inverses, but a nonsingular matrix does. One means of constructing it is as follows.

Let $A = (a_{ij})$ be a nonsingular matrix. The *cofactor matrix* of A is the matrix

$$\mathrm{co}(A) = (A_{ij})$$

obtained by replacing each element of A with its cofactor. The transpose of the cofactor matrix is called the *adjoint* of A

$$\mathrm{adj}(A) = \mathrm{co}(A)^{\mathrm{T}}.$$

If we let $A' = \mathrm{adj}(A)$, then by properties of determinants, we have

$$AA' = \begin{pmatrix} a_{11} & a_{12} & \cdots & a_{1n} \\ a_{21} & a_{22} & \cdots & a_{2n} \\ \vdots & \vdots & & \vdots \\ a_{n1} & a_{n2} & \cdots & a_{nn} \end{pmatrix} \begin{pmatrix} A_{11} & A_{21} & \cdots & A_{n1} \\ A_{12} & A_{22} & \cdots & A_{n2} \\ \vdots & \vdots & & \vdots \\ A_{1n} & A_{2n} & \cdots & A_{nn} \end{pmatrix}$$

$$= \begin{pmatrix} |A| & 0 & \cdots & 0 \\ 0 & |A| & \cdots & 0 \\ \vdots & \vdots & & \vdots \\ 0 & 0 & \cdots & |A| \end{pmatrix} = |A|I.$$

We also have $A'A = |A|I$. Hence

$$A^{-1} = \frac{1}{|A|}\ \mathrm{adj}(A)$$

is the inverse of A.

A singular matrix has no inverse, for if A is singular and B is its inverse, then $AB = I$ and

$$1 = |I| = |AB| = |A||B| = 0|B| = 0,$$

an absurdity.

Systems of Linear Equations

A *system of linear equations* is a set of m equations

$$\begin{aligned} a_{11}x_1 + a_{12}x_2 + \cdots + a_{1n}x_n &= b_1, \\ a_{21}x_1 + a_{22}x_2 + \cdots + a_{2n}x_n &= b_2, \\ &\vdots \\ a_{m1}x_1 + a_{m2}x_2 + \cdots + a_{mn}x_n &= b_m \end{aligned}$$

in the n unknowns $x_1, \ldots, x_n$. If

$$A = \begin{bmatrix} a_n & a_{12} & \cdots & a_{1n} \\ a_{21} & a_{22} & \cdots & a_{2n} \\ \vdots & \vdots & & \vdots \\ a_{m1} & a_{m2} & \cdots & a_{mn} \end{bmatrix}, \qquad X = \begin{bmatrix} x_1 \\ x_2 \\ \vdots \\ x_n \end{bmatrix}, \qquad B = \begin{bmatrix} b_1 \\ b_2 \\ \vdots \\ b_m \end{bmatrix},$$

then the system can be written in the matrix form $AX = B$.

A *solution* of the system is a column vector S of length n such that

$$AS = B.$$

A system that has no solution is called *inconsistent*; one that has more than one solution is called *dependent*.

The matrix A is called the *coefficient matrix* of the system $AX = B$, and B is called the *constant column*. The matrix

$$C = \begin{bmatrix} a_{11} & a_{12} & \cdots & a_{1n} & b_1 \\ a_{21} & a_{22} & \cdots & a_{2n} & b_2 \\ \vdots & \vdots & & \vdots & \vdots \\ a_{m1} & a_{m2} & \cdots & a_{mn} & b_n \end{bmatrix}$$

is called the *augmented matrix* of the system; C is also denoted by $(A \mid B)$.

The basic result on systems of linear equations is that the system has a solution if and only if the coefficient matrix and the augmented matrix have the same rank.

The finding of solutions depends somewhat on the form of the system. One general method is the Gauss reduction method, which we shall not discuss here. Since the systems with which we are concerned have certain special forms, we shall discuss methods more easily applied to those forms.

A *square* system is one in which the coefficient matrix is square, that is, in which the number of equations equals the number of unknowns. If the coefficient matrix is nonsingular, then the system $AX = B$ has the unique solution

$$X = A^{-1}B.$$

If the coefficient matrix is singular, the system is either inconsistent or dependent. If the system is consistent, that is, has a solution, and the coefficient matrix has rank r, then the solution of the system is found by solving r of the equations for r of the unknowns in terms of the remaining unknowns. Arbitrary values may be assigned to the remaining unknowns; thus if there were originally n unknowns, there will be $n - r$ degrees of freedom in the solution.

Another method of finding the solution of a square system with nonsingular coefficient matrix is *Cramer's rule*. Let $D = |A|$, and let D_j be the determinant of the matrix obtained from A by replacing the jth column of A with the constant column. Then $X = (x_1, \ldots, x_n)^{\mathrm{T}}$ is found by the formula

$$x_j = \frac{D_j}{D}.$$

Cramer's rule is not usually any more efficient, but it has the convenience of allowing for a solution for any one of the unknowns without having to find solutions for all of them.

A *homogeneous* system is one in which the constant column consists of zeros. A homogeneous system is denoted by $AX = 0$. Since $(A \mid 0)$ has the same rank as A, a homogeneous system always has a solution. In fact, the zero vector $X = (0, \ldots, 0)^{\mathrm{T}}$ is always a solution, called the *trivial* solution. Thus in a homogeneous system the question is whether there exists a nontrivial solution.

A square homogeneous system with nonsingular coefficient matrix has only the trivial solution, since $A^{-1}0$ is the zero vector. Thus a square homogeneous system has a nontrivial solution if and only if the coefficient matrix A is singular, or $|A| = 0$.

Finding the nontrivial solutions of a homogeneous system is the same as finding the solutions of a dependent, consistent system. In particular, if a homogeneous system of $n - 1$ equations in n unknowns has coefficient matrix of rank $n - 1$, let D_j be the subdeterminant of A obtained by deleting the jth column of A. Then the solutions $X = (x_1, \ldots, x_n)^{\mathrm{T}}$ are found by the formula

$$x_j = (-1)^j k D_j,$$

k an arbitrary constant.

Mappings

Let S and T be sets. A *mapping* from S to T is a function with domain S and range in T. The notation

$$f : S \to T$$

indicates that f is a mapping from S to T. For $x \in S$, the element $y = f(x) \in T$ is called the *image* of x, and x is called a *preimage* of y. The notation

$$f : x \to y$$

is used to indicate that x has image y. If U is a subset of S and $f: S \to T$, the set

$$f(U) = \{ f(x) \mid x \in U \}$$

is called the *image of* U. The image of S is of course the range of f. If V is a subset of T, the set

$$f^{-1}(V) = \{ x \in S \mid f(x) \in V \}$$

is called the *counterimage of* V. The counterimage of T is the domain of f. If $y \in T$, the counterimage of y is thus the set of all preimages of y.

The *identity mapping* on set S is the function $i : S \to S$ defined by $i(x) = x$ for each $x \in S$.

A mapping $f : S \to T$ is *one-to-one* or *injective* if and only if any one of the following equivalent statements holds:

1. If $x_1, x_2 \in S$ and $f(x_1) = f(x_2)$, then $x_1 = x_2$.
2. If $x_1, x_2 \in S$ and $x_1 \neq x_2$, then $f(x_1) \neq f(x_2)$.
3. If $y \in f(S)$, then $f^{-1}(y)$ has exactly one element.
4. If $y \in T$, then $f^{-1}(y)$ has at most one element.

A mapping $f: S \to T$ is *onto* or *surjective* if and only if any one of the following equivalent statements holds:

1. If $y \in T$, then $f(x) = y$ for some $x \in S$.
2. If $y \in T$, then $f^{-1}(y)$ has at least one element.
3. $f(S) = T$.

A mapping is *bijective* if and only if it is both injective and surjective. A mapping $f: S \to T$ that is bijective has an *inverse mapping* $f^{-1} : T \to S$ that is also bijective, and having the properties $f^{-1}(f(x)) = x$ for each $x \in S$, and $f(f^{-1}(y)) = y$ for each $y \in T$. An identity mapping is bijective.

If $f: S \to T$ and U is a subset of S, the mapping

$$f|_U : U \to T$$

is called a *restriction* of f, and is defined by

$$f|_U(x) = f(x)$$

for each $x \in U$. That is, $f|_U$ is simply f, restricted to a smaller domain.

If $f : S \to T$ and $g : T \to U$ are mappings, the *composition* of f and g is the mapping $g \circ f : S \to U$ defined by $(g \circ f)(x) = g(f(x))$. The composition of f and g is sometimes denoted by gf. A composition of injective (surjective, bijective) mappings is injective (surjective, bijective). The composition of a bijective function and its inverse is an identity mapping.

Relations

Let S and T be sets. A *relation* from S to T is a subset of

$$S \times T = \{(x, y) | x \in S \text{ and } y \in T\}.$$

If R is a relation from S to T and $(x, y) \in R$, we write $x R y$.

A relation R from S to S is called a *relation on* S. R is *reflexive* if and only if $x R x$ for each $x \in S$. R is *symmetric* if and only if $x R y$ implies $y R x$. R is *transitive* if and only if $x R y$ and $y R z$ imply $x R z$. A relation on S which is reflexive, symmetric, and transitive is called an *equivalence relation*. If R is an equivalence relation on S and $x \in S$, the *equivalence class* of x is the set

$$[x] = \{y \in S | x R y\}.$$

If $x R y$, then $[x] = [y]$, and if $x R y$, then $[x] \cap [y] = \emptyset$. Thus each element of S belongs to exactly one equivalence class: the equivalence classes form a *partition* of S.

Operations

Let S be a set. A *binary operation* on S is a mapping

$$\circ : S \times S \to S.$$

If $(x, y) \in S \times S$ has image z under the operation $\circ$, we write

$$x \circ y = z.$$

A binary operation $\circ$ on S is *associative* if and only if $(x \circ y) \circ z = x \circ (y \circ z)$ for all $x, y, z \in S$; it is *commutative* if and only if $x \circ y = y \circ x$ for all $x, y \in S$. An element $e \in S$ is an *identity* for $\circ$ if and only if $e \circ x = x \circ e = x$ for all $x \in S$. If an identity exists for $\circ$, it is unique. If $\circ$ has an identity e, an element $x \in S$ has *inverse* x' relative to $\circ$ if and only if $x \circ x' = x' \circ x = e$.

If $\circ$ is a binary operation on S and U is a subset of S, then U is *closed under* $\circ$ if and only if $x, y \in U$ implies $x \circ y \in U$.

If $\circ$ and $*$ are two binary operations on S, $\circ$ is *left-distributive over* $*$ if and only if $x \circ (y * z) = (x \circ y) * (x \circ z)$ for all $x, y, z \in S$; $\circ$ is *right-distributive* over $*$ if and only if $(x * y) \circ z = (x \circ z) * (y \circ z)$ for all $x, y, z \in S$. If $\circ$ is both left- and right-distributive over $*$, we say $\circ$ is *distributive* over $*$.

The common binary operations, addition (+) and multiplication (·), have their own notations, as shown in the following table:

	Neutral	Addition	Multiplication
Operation	$\circ$	$+$	$\cdot$
Identity	e	0	1
Inverse	x'	$-x$	x^{-1}

Groups

A *group* is a set G, together with a binary operation $\circ$ on G, satisfying

1. $\circ$ is associative,
2. $\circ$ has an identity e,
3. for each $x \in G$, there exists an inverse $x' \in G$ relative to $\circ$.

If the operation on G is commutative, we call G an *abelian* group.

A *subgroup* of a group G is a subset S of G that forms a group itself, under the operation on G. Thus $S \subseteq G$ will be a subgroup of G if and only if

1. S is closed under $\circ$ and
2. for each $x \in S$, $x' \in S$.

Rings

A *ring* is a set R, together with two binary operations, addition (+) and multiplication (· or juxtaposition), such that

1. R is an abelian group under $+$,
2. $\cdot$ is associative,
3. $\cdot$ is distributive over $+$.

If $\cdot$ is commutative, R is called a *commutative ring*. If $\cdot$ has an identity 1, then 1 is called the *unity* of R. Any element x of R that has an inverse x^{-1} relative to $\cdot$ is called a *unit* of R.

The most familiar example of a ring is the ring of integers, denoted by Z. The rational numbers Q, the real numbers R, and the complex numbers C also form rings.

Other examples of rings are the matrix ring $M_n(R)$ of $n \times n$ matrices with real entries, the polynomial ring $R[x]$ of all polynomials with real coefficients, and the ring $R(x)$ of all rational functions in x with real coefficients.

Division Rings and Fields

A *division* ring is a ring D whose nonzero elements D^* form a group under $\cdot$. A commutative division ring is a *field*.

The simplest example of a division ring that is not a field is the ring of *real quaternions* $\mathcal{Q}$, which we now describe. Over the field of complex numbers C, we form all 2×2 matrices of the form

$$\begin{pmatrix} a & b \\ -\bar{b} & \bar{a} \end{pmatrix},$$

with $a, b \in C$, and where $\bar{a}$ stands for the complex conjugate of a. It is easily verified that the sum and product of two matrices of that form is another matrix of that form, the negative of such a matrix is another such matrix, and the inverse of such a matrix,

$$\begin{pmatrix} a & b \\ -\bar{b} & \bar{a} \end{pmatrix}^{-1} = \frac{1}{a\bar{a} + b\bar{b}} \begin{pmatrix} \bar{a} & -b \\ \bar{b} & a \end{pmatrix},$$

is another matrix of that form. It follows that the set $\mathcal{Q}$ of such matrices is a division ring under matrix addition and multiplication. But multiplication is not commutative, as the following shows:

$$\begin{pmatrix} 0 & 1 \\ -1 & 0 \end{pmatrix}\begin{pmatrix} i & 0 \\ 0 & -i \end{pmatrix} = \begin{pmatrix} 0 & -i \\ -i & 0 \end{pmatrix},$$

$$\begin{pmatrix} i & 0 \\ 0 & -i \end{pmatrix}\begin{pmatrix} 0 & 1 \\ -1 & 0 \end{pmatrix} = \begin{pmatrix} 0 & i \\ i & 0 \end{pmatrix}.$$

The quaternions $\mathcal{Q}$ are often given in a different form. If $a = a_1 + ia_2$, $b = b_1 + ib_2$, then

$$\begin{pmatrix} a & b \\ -\bar{b} & \bar{a} \end{pmatrix} = \begin{pmatrix} a_1 + ia_2 & b_1 + ib_2 \\ -b_1 + ib_2 & a_1 - ia_2 \end{pmatrix}$$

$$= a_1\begin{pmatrix} 1 & 0 \\ 0 & 1 \end{pmatrix} + a_2\begin{pmatrix} i & 0 \\ 0 & -i \end{pmatrix} + b_1\begin{pmatrix} 0 & 1 \\ -1 & 0 \end{pmatrix} + b_2\begin{pmatrix} 0 & i \\ i & 0 \end{pmatrix}.$$

If we set

$$\mathbf{i} = \begin{pmatrix} i & 0 \\ 0 & -i \end{pmatrix},$$

$$\mathbf{j} = \begin{pmatrix} 0 & 1 \\ -1 & 0 \end{pmatrix},$$

$$\mathbf{k} = \begin{pmatrix} 0 & i \\ i & 0 \end{pmatrix},$$

then

$$\begin{pmatrix} a & b \\ -\bar{b} & \bar{a} \end{pmatrix} = a_1 + a_2\mathbf{i} + b_1\mathbf{j} + b_2\mathbf{k}.$$

Thus $\mathcal{Q}$ can be described as the set of all objects of the form

$$a + b\mathbf{i} + c\mathbf{j} + d\mathbf{k},$$

with $a, b, c, d \in R$, under the regular addition and multiplication, using the rules

$$\mathbf{i}^2 = \mathbf{j}^2 = \mathbf{k}^2 = -1,$$
$$\mathbf{ij} = \mathbf{k}, \quad \mathbf{jk} = \mathbf{i}, \quad \mathbf{ki} = \mathbf{j},$$
$$\mathbf{ji} = -\mathbf{k}, \quad \mathbf{kj} = -\mathbf{i}, \quad \mathbf{ik} = -\mathbf{j}.$$

The familiar number systems Q, R, and C are all fields. If F is a field, then the ring $F(x)$ of rational functions in x with coefficients in F is also a field.

Another class of fields should be mentioned here, the *finite* fields. The simplest of these are *modular systems*, described as follows.

Let p be a positive integer that is prime. Let the relation $\equiv$ be defined on Z by $a \equiv b$ if and only if $a - b$ is divisible by p. Then $\equiv$ is an equivalence relation on Z, called *congruence modulo p*. Let $[a]$ be the equivalence class of a under $\equiv$, and let Z_p be the set of all equivalence classes. Then Z_p has exactly p elements:

$$Zp = \{[0], [1], [2], \ldots, [p-1]\}.$$

Let $+$ and $\cdot$ be defined on Z_p by

$$[a] + [b] = [a + b],$$
$$[a] \cdot [b] = [ab].$$

Then $+$ and $\cdot$ are well-defined binary operations, and Z_p is a field under $+$ and $\cdot$. Z_p is called the field of *integers modulo p*.

The *order* of a finite field is the number of elements in the field; hence the order of Z_p is p.

A *subfield* F of a field K is a subset of K that is a field under the operations of K. Thus $F \subseteq K$ is a subfield of K if and only if

1. F is closed under $+$ and $\cdot$,
2. if $x \in F$, then $-x \in F$, and
3. if $x \in F \backslash \{0\}$, then $x^{-1} \in F$.

Vector Spaces

Let D be a division ring. A *left vector space over D* is a set V of elements called *vectors*, together with a binary operation $\oplus$ on V and a binary operation $D \times V \to V$, denoted by juxtaposition $[(a, v) \to av]$, such that

1. V is an abelian group under $\oplus$,
2. $a(bv) = (ab)v$ for $a, b \in D$, $v \in V$,
3. $(a + b)v = av \oplus bv$ for $a, b \in D$, $v \in V$,

4. $a(u \oplus v) = au \oplus av$ for $a \in D$, $u, v \in V$, and
5. $1v = v$.

The operation $\oplus$ is called *vector addition*, and the operation $(a, v) \rightarrow av$ is called *scalar multiplication*. The identity in V relative to $+$ is called the *zero vector*.

The simplest examples of vector spaces over D are the *tuple spaces*. Let

$$D^n = D \times D \times \cdots \times D$$

be the n-fold Cartesian product of D with itself, consisting of n-tuples of the form

$$(x_1, x_2, \ldots, x_n),$$

for each $x_i \in D$. Let vector addition be defined elementwise, as for matrices,

$$(x_1, x_2, \ldots, x_n) \oplus (y_1, y_2, \ldots, y_n) = (x_1 + y_1, x_2 + y_2, \ldots, x_n + y_n),$$

and let scalar multiplication also be defined as for matrices,

$$a(x_1, x_2, \ldots, x_n) = (ax_1, ax_2, \ldots, ax_n).$$

Then D^n is a left vector space over D, and is of *dimension n*. It can be shown that any n-dimensional left vector space over D is isomorphic to D^n.

Solutions to Selected Exercises

Exercises 1.1

1. **(a)** Yes. **(b)** Yes. **(c)** B. **(d)** Yes. **(e)** Yes. **(f)** No. **(g)** a. **(h)** a, b. **(i)** A, B.

2. 2^{mn}.

Exercises 1.2

3. **(a)** is A1 and **(b)** is A2. Let p, q, r be the three noncollinear points. Then pq is a line, so A5 holds. Not all of p, q, r are on any line, so A4 holds. Finally, if L is any line, one of p, q, r is not on L, say p. If pq and pr both meet L, then L has two points. If pq, say, is parallel to L, then pr meets L. Let M be the line on q parallel to pr; then M meets L, and L has two points. Hence A3 holds.

5. **(a)** is P1 and **(b)** is P2. Let a, b, c, d be the four-point. Then ab is a line, and not all of a, b, c are on any line, so P5 and P4 hold. For any line L, suppose a is not on L. Then ab, ac, ad meet L in three points, and P3 holds.

Exercises 1.3

1. If, for instance, $a \neq 0$ and $b = 0$, then points $(-c/a, 0)$ and $(-c/a, y)$ are on the line; if $a \neq 0$ and $b \neq 0$, then $(-c/a, 0)$ and $(0, -c/b)$ are on the line. $(1, 0)$ is not on the line unless $a + c = 0$; then $(2, 0)$ is not.

3. **(b)** $a_0a_1 = \langle -1, -1, 1\rangle$; $a_2a_3 = \langle 1, -1, 0\rangle$. **(c)** $d_1 = [1, 1, 2]$.

7. **(a)** $(1, 1)$. **(c)** $(1, 0)$. **(e)** $(2, 1)$. **(g)** $(4, 5)$. **(i)** $(\sqrt{2}, -1)$.

10. We have

$$c_1 = \lambda a_1 + \mu b_1 = \lambda' a_1' + \mu' b_1',$$
$$c_2 = \lambda a_2 + \mu b_2 = \lambda' a_2' + \mu' b_2',$$
$$c_3 = \lambda a_3 + \mu b_3 = \lambda' a_3' + \mu' b_3'.$$

Since the matrix

$$\begin{pmatrix} a_1' & a_2' & a_3' \\ b_1' & b_2' & b_3' \end{pmatrix}$$

has rank 2, suppose

$$\begin{vmatrix} a_2' & a_2' \\ b_1' & b_2' \end{vmatrix} \neq 0.$$

Then also

$$\begin{vmatrix} a_1 & a_2 \\ b_1 & b_2 \end{vmatrix} \neq 0,$$

for both are first coordinate of the line. Solving the first two equations above for λ' and μ', we get

$$\lambda' = \lambda\begin{vmatrix} a_1 & b_1' \\ a_2 & b_2' \end{vmatrix} + \mu\begin{vmatrix} b_1 & b_1' \\ b_2 & b_2' \end{vmatrix},$$

$$\mu' = \lambda\begin{vmatrix} a_1' & a_1 \\ a_2' & a_2 \end{vmatrix} + \mu\begin{vmatrix} a_1' & b_1 \\ a_2' & b_2 \end{vmatrix};$$

let

$$M = \begin{pmatrix} \begin{vmatrix} a_1 & b_1' \\ a_2 & b_2' \end{vmatrix} & \begin{vmatrix} a_1' & a_1 \\ a_2' & a_2 \end{vmatrix} \\ \begin{vmatrix} b_1 & b_1' \\ b_2 & b_2' \end{vmatrix} & \begin{vmatrix} a_1' & b_1 \\ a_2' & b_2 \end{vmatrix} \end{pmatrix}.$$

Note that

$$|M| = \begin{vmatrix} a_1 & b_1 \\ a_2 & b_2 \end{vmatrix}\begin{vmatrix} a_1' & b_1' \\ a_2' & b_2' \end{vmatrix} \neq 0.$$

Exercises 1.4

2. Yes, but the proof is difficult. See Garner [10].

Exercises 1.5

2. No.

3. Yes. Let $\sigma \in C$. Then S is true of σ, so S^d is true of σ. Therefore S is true of σ^d, and $\sigma^d \in C$.

Exercises 1.6

1. Triangle: Figure 1.1. Complete four-point: Figure S.1. Fano configuration: Figure 1.14. Pappus configuration: Figure 3.26. Desargues configuration: Figure 3.3.

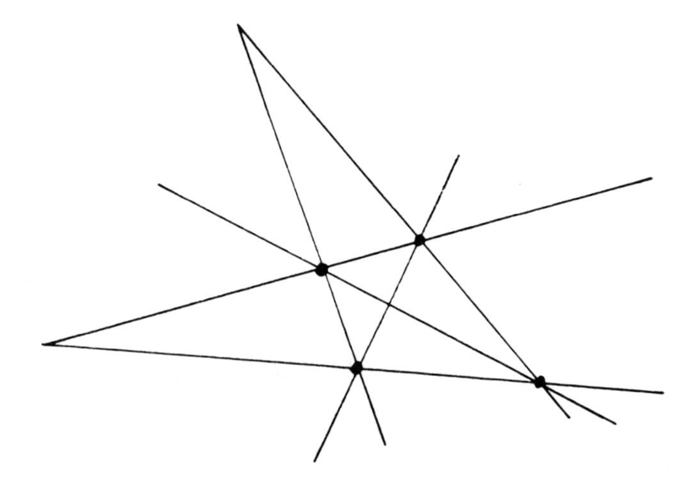

Figure S.1

3. All but the complete four-point.

7. **(c)** Figure 1.15. **(e)** Figure S.2.

9. See Exercise 7(b).

10. 21.

Exercises 1.7

1. Use Exercise 1.2.4.

Figure S.2

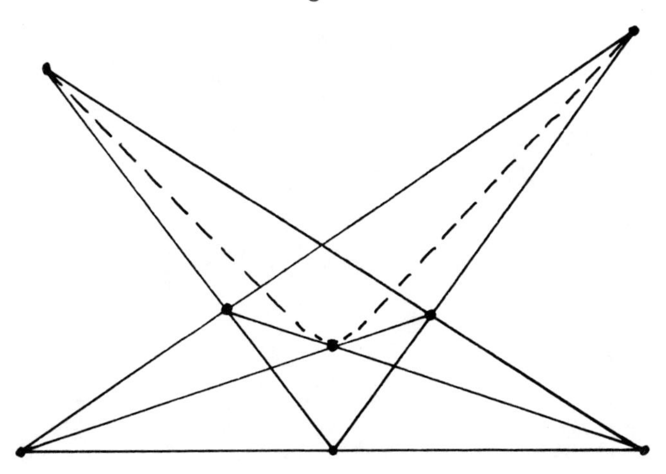

3. For example, let $a = [1, 0, 1]$, $b = [1, 1, 1]$, $c = [0, 1, 1]$, and $a' = [2, 0, 1]$, $b' = [2, 2, 1]$, $c' = [0, 2, 1]$. Compute the remaining points and lines.

4. **(a)** $x_1 + 3x_2 = 5x_3$. **(c)** $x_1x_2 - x_3^2$. **(e)** $x_1x_2 + x_1^2 - 2x_3^2 = 0$.

5. **(a)** $[3, -1, 0]$. **(c)** $[1, 0, 0]$, $[0, 1, 0]$. **(e)** $[0, 1, 0]$, $[-1, 1, 0]$.

6. Use the discriminant of the conic section.

Exercises 1.8

2. An isomorphism is shown in Figure S.3 by labeling the points.

5. First we show at most one line is on two given points. Given points $[x_1, x_2, 1]$ and $[y_1, y_2, 1]$, if $x_1 = y_1$, then only line $\langle 1, 0, -x_1\rangle$ is on both. If $x_1 \neq y_1$, then only line $\langle -(x_1 - y_1)^{-1}(x_2 - y_2), 1, x_1(x_1 - y_1)^{-1}(x_2 - y_2) - x_2\rangle$ is on both. Given points $[x_1, x_2, 1]$ and $[0, y_2, 0]$, only line $\langle -y_2, 1, x_1y_2 - x_2\rangle$ is on both. Given points $[x_1, x_2, 1]$ and $[0, 1, 0]$, only line $\langle 1, 0, -x_1\rangle$ is on both. Given points $[1, x_2, 0]$ and $[1, y_2, 0]$, only line $\langle 0, 0, 1\rangle$ is on both. Given points $[1, x_2, 0]$ and $[0, 1, 0]$, only line $\langle 0, 0, 1\rangle$ is on both. Next we show each line has two points on it. Line $\langle l_1, 1, l_3\rangle$ has points $[1, -l_1, 0]$ and $[0, -l_3, 1]$ on it. Line $\langle 1, 0, l_3\rangle$ has points $[-l_3, 0, 1]$ and $[0, 1, 0]$ on it. Line $\langle 0, 0, 1\rangle$ has points $[1, 1, 0]$ and $[1, 0, 0]$ on it.

8. No.

Figure S.3

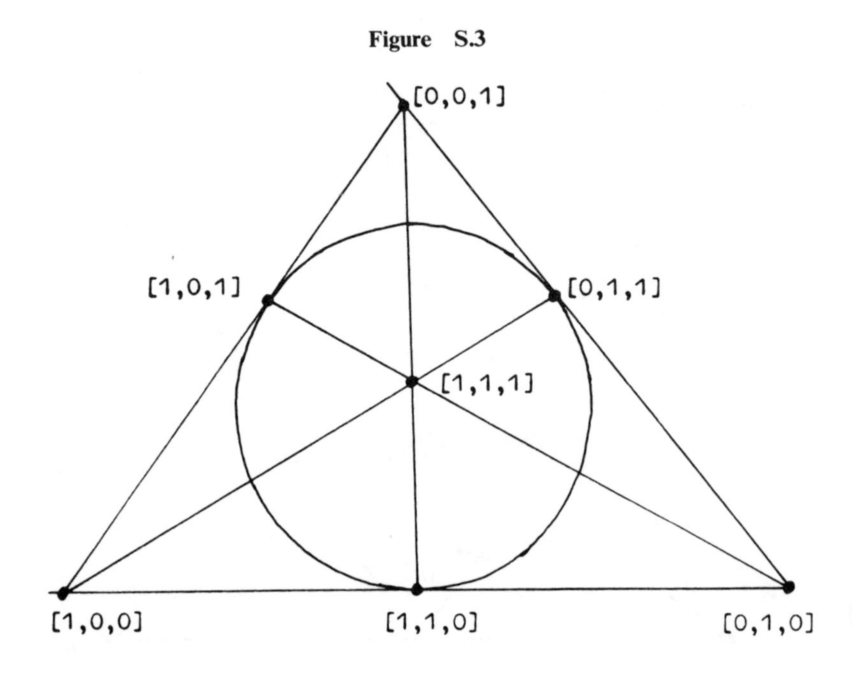

9. No; π_V is of that form.

10. $((n-1)^2_n, n^2 - n_{n-1})$.

11. $(n^2 + n + 1_n)$.

14. No one knows. See Dembowski [7, p. 144].

Exercises 2.1

1. If p, q are on L and p', q' are their images on L', and $p \neq p'$ and $q \neq q'$, then the center $u = pp' \cap qq'$ is uniquely determined. However, if $p = p' = L \cap L'$ or $q = q' = L \cap L'$, the center is not uniquely determined.

3. $\lambda\mu' + 2\mu\lambda' = 0$, or $(\lambda', \mu') = (\lambda, \mu)\begin{pmatrix} -1 & 0 \\ 0 & 2 \end{pmatrix}$.

5. $v = [-2, 0, 1]$.

Exercises 2.2

2. Use Theorem 2.1.1.

3. Let L_1 be a distinct line, and let $g : L(abc) \barwedge L_1(a_1b_1c_1)$ be any convenient perspectivity. Then use the case proved to construct a sequence of two perspectivities from L_1 to L' $(= L)$.

4. $[4, 2, -2]$, $[2, 1, 1]$ $[4, 2, 3]$, $[6, 3, 7]$.

5. $-2\lambda\lambda' - 3\lambda\mu' + \mu\lambda' + \mu\mu' = 0$, or $(\lambda', \mu') = (\lambda, \mu)\begin{pmatrix} -3 & 2 \\ 1 & -1 \end{pmatrix}$.

7. $f(q) = f(f(p)) = f^2(p) = p$.

9. If

$$(\lambda, \mu)\begin{pmatrix} a & b \\ c & d \end{pmatrix} = k(\lambda, \mu),$$

then

$$(\lambda, \mu)\left[\begin{pmatrix} a & b \\ c & d \end{pmatrix} - \begin{pmatrix} k & 0 \\ 0 & k \end{pmatrix}\right] = (0, 0),$$

which is the homogeneous system

$$(a-k)\lambda + c\mu = 0,$$

$$b\lambda + (d-k)\mu = 0.$$

This system has a nontrivial solution if and only if

$$\begin{vmatrix} a-k & c \\ b & d-k \end{vmatrix} = 0.$$

10. (1, 1) and (1, − 1).

12. Zero, one, or two, depending on how many solutions for k the equation

$$\begin{vmatrix} a-k & c \\ b & d-k \end{vmatrix} = 0$$

has in F.

Exercises 2.3

4. Function composition is associative, the identity is a collineation, and the inverse of an isomorphism is an isomorphism. Moreover, the composition of isomorphisms is an isomorphism, so $C(\pi)$ has all the properties of a group.

5. Yes. We denote it by $P(\pi)$.

Exercises 2.4

2. **(a)** [0, 1, 1] and [0, 0, 1]. **(b)** $[-2, -3, 1]$ and every point on line $\langle -1, 2, 3 \rangle$.

6. The system is

$$h_1x_1 + h_2y_1 + h_3z_1 = p_1,$$

$$h_1x_2 + h_2y_2 + h_3z_2 = p_2,$$

$$h_1x_3 + h_2y_3 + h_3z_3 = p_3.$$

Point $[p_1, p_2, p_3]$ can lie on no side of triangle xyz.

7.

(a) $M = \begin{bmatrix} \frac{8}{7} & -\frac{4}{7} & 0 \\ \frac{6}{7} & -\frac{2}{7} & \frac{2}{7} \\ 0 & \frac{9}{7} & -\frac{9}{7} \end{bmatrix}$. **(b)** $M = \begin{pmatrix} 68 & 49 & 23 \\ -68 & -70 & -2 \\ -20 & -1 & 1 \end{pmatrix}$.

8. If $p_1p_2p_3p_4$ and $q_1q_2q_3q_4$ are the four-point and its image, and M is a 3×3 matrix, the four matrix equations

$$p_iM = k_iq_i \qquad (i = 1, \ldots, 4)$$

become twelve linear equations in the nine entries of M and the four constants k_i, so a solution exists.

Exercises 2.5

1. Use Theorem 2.3.1

3. If $a = px \cap A$, then $x' = xc \cap ap'$.

4. Yes, if pp' and qq' are distinct lines.

5. Yes, if pqr and $p'q'r'$ are the triangles of a Desargues configuration.

7. No. The composition of central collineations need not be a central collineation.

Exercises 2.6

1. Let p be an arbitrary point, and let q be a point on the vanishing line V such that pq is not on the center c. Let pq meet the axis A at r. Let L be the line on r parallel to cq. Then $p' = pc \cap L$.

4. Let the triangle be half of a quadrilateral, and transform the quadrilateral into a square.

6. If the circle meets the vanishing line in a certain number of points, then its image meets the ideal line in the same number of points. Now use Exercise 1.7.6.

7. The image of (a, b) is

$$\left(\frac{3a}{a+1}, \frac{a+3b-2}{a+1} \right).$$

If the image of (a, b) is (a', b'), then

$$a = \frac{a'}{3-a'} \quad \text{and} \quad b = \frac{3b'-2a'+6}{3(3-a')}.$$

The image of $x^2 + y^2 = 1$ is the parabola $4x'^2 - 12x'y' + 9y'^2 + 30x' + 36y' - 45 = 0$.

8. Let c be the center of the plane perspective, and let t be the point common to the circle and the vanishing line V. Let the line perpendicular to ct at c meet V at u, and let v be the point of contact of the second tangent to the circle from u. Then v becomes the vertex.

9. Let c be the center of the plane perspective, and let the vanishing line V meet the circle at t_1 and t_2. Let the tangents to the circle at t_1 and t_2 meet at p. Let q be the point between t_1 and t_2 such that cq bisects angle t_1ct_2. Then pq meets the circle in the required points.

10. Use Theorem 2.

11. Regard a circle as the locus of points forming a right angle with the endpoints of a diameter, and use Exercise 10.

12. If s, s' are another finite point and its image, then $\triangle psc$ is similar to $\triangle p's'c'$, so that $(s'c)/(sc) = (p'c)/(pc) = r$.

15. Let r be the ratio of homothecy f. Then f is harmonic iff f^2 is the identity, iff $r^2 = 1$. If $r = 1$, f is the identity; hence $r = -1$ if f is harmonic. Conversely, $r = -1$ implies $r^2 = 1$, so f is harmonic.

16. The ratios must be 1 or -1.

Exercises 2.7

1. Yes.

4. Let h_1 be the identity, and $h_2 = g^{-1}$.

Exercises 3.1

2. Use Theorem 2 and Exercise 1.5.3.

4. The four-point $[1, 1, 1]$, $[1, 0, 0]$, $[0, 1, 0]$, $[0, 0, 1]$ has noncollinear diagonal points, while the four-point $[0, 0, 1]$, $[i, 0, 1]$, $[2k, 1, 1]$, $[1, j, 0]$ has collinear diagonal points. See further Stevenson [16, p. 81].

6. Points p, r, and t must be distinct; if q is not also distinct, then $q = p$; if s is not distinct, then $s = r$.

8. Let a_0 be any point not on gh, and let a_1 be a third point on a_0g. Let $a_2 = a_0h \cap a_1x$ and $a_3 = a_1h \cap a_2g$. Then $y = a_0a_3 \cap gh$ is the harmonic conjugate of x relative to g and h.

10. (d_1d_2, h_3h_3') is a harmonic set, induced on d_1d_2 by the four-point $a_3a_2a_1a_0$; the other cases are similar.

11. For h_1', h_2', h_3', use the couples $a_1a_2a_3$ and $d_1d_2d_3$; for h_1', h_2, h_3, use couples $a_0a_2a_3$ and $d_1d_3d_2$; etc.

12. Use Exercise 11.

13. π_R is Desarguesian, and satisfies Fano's axiom by Exercise 1.3.3. Hence the Fano configuration cannot lie in π_R, nor in α_R, which is a subset of π_R.

Exercises 3.2

1. Take one perspectivity to get to a different range, and then at most two (as per Theorem 4) to get back on.

3. Let M be a second line on p, and let u be a point not on L or M. Let $s = qu \cap M$, $t = ru \cap M$, and $v = rs \cap qt$. Then $L(pqr) \overset{u}{\barwedge} M(pst) \overset{v}{\barwedge} L(prq)$.

5. We cannot tell yet. It is if three points and their images determine a projectivity uniquely.

Exercises 3.3

3. See Figures 2.10 and 2.11.

4. A projective collineation carries the four-point [1, 0, 0], [0, 1, 0], [0, 0, 1], [1, 1, 1] to another four-point $uvoe$, which is then the reference quadrangle of the new coordinate system. See further Ayers [3, p. 179].

Exercises 3.4

1. Prove that the dual of the theorem of Pappus is a consequence of the theorem of Pappus, and use Exercise 1.5.3.

3. Let $h = g \circ f$.

6. Use Theorems 3.3.3 and 3.4.1.

7. Imitate Exercise 1.5.4.

9. Let a be a point on L and let a' be a point on L' but not on ap. Let $b' = ap \cap L'$, $b = a'p \cap L$, and $u = aa' \cap bb'$. Let c be a third point on L, and let $c' = uc \cap L'$, and $q = bc' \cap b'c$. Then line pq is on $L \cap L'$, for triangles $ab'c$ and $a'bc'$ form a central (hence axial) couple.

10. Let x be any point on L, and let $f(x) = x'$. Then $f\colon L(abx) \sim L(bax')$. But by the permutation theorem, there is a projectivity $g : L(abxx') - L(bax'x)$, so by the fundamental theorem $f = g$, and $f^2(x) = f(f(x)) = f(x') = x$.

Exercises 3.5

1. **(a)** $\frac{4}{3}$. **(b)** -4.

2. If there are exactly three distinct points among p_1, p_2, p_3, p_4, then $\mathcal{R}(p_1p_2p_3p_4)$ is 0 when $p_1 = p_3$ or $p_2 = p_4$, 1 when $p_1 = p_2$ or $p_3 = p_4$, and undefined when $p_1 = p_4$ or $p_2 = p_3$. If there are exactly two points, the cross ratio is 0 if $p_1 = p_3$ and $p_2 = p_4$, 1 if $p_1 = p_2$ and $p_3 = p_4$, and undefined otherwise. The cross ratio of only one point is not defined.

5. Use the permutation theorem, Theorem 3.4.8.

6. The values are $r, 1/r, 1-r, 1/(1-r), (r-1)/r, r/(r-1)$.

8. One of the determinants

$$\begin{vmatrix} x_1 & x_2 \\ y_1 & y_2 \end{vmatrix}, \quad \begin{vmatrix} x_2 & x_3 \\ y_2 & y_3 \end{vmatrix}, \quad \begin{vmatrix} x_3 & x_1 \\ y_3 & y_1 \end{vmatrix}$$

is not zero; then get parameters for p_3 and p_4 relative to p_1 and p_2, and compute the cross ratio by brute force.

10. **(a)** p. **(b)** q.

12. First let T be perpendicular to L_3, and use Exercise 11. Then use Theorem 3 for any transversal.

14. Let c_1r_1 and c_2r_2 be parallel radii of the two circles, and let c_2r_2' be the radius of γ_2 opposite to c_2r_2 (so that r_2r_2' is a diameter of γ_2). Let r_1r_2 and r_1r_2' meet c_1c_2 at t and t'. Then the tangents to γ_1 from t and t' are common tangents.

16. (ab, cd) is a harmonic set

$$\Leftrightarrow \frac{(ac)}{(cb)} = -\frac{(ad)}{(db)}$$

$$\Leftrightarrow \frac{(bc)}{(ac)} = -\frac{(bd)}{(ad)}$$

$$\Leftrightarrow \frac{(ba)+(ac)}{(ac)} = -\frac{(ba)+(ad)}{(ad)}$$

$$\Leftrightarrow \frac{(ba)}{(ac)} + 1 = -\frac{(ba)}{(ad)} - 1$$

$$\Leftrightarrow \frac{1}{(ac)} + \frac{1}{(ba)} = -\frac{1}{(ad)} - \frac{1}{(ba)}$$

$$\Leftrightarrow \frac{1}{(ac)} + \frac{1}{(ad)} = \frac{2}{(ab)},$$

$$\Leftrightarrow \frac{1}{(ab)} - \frac{1}{(ac)} = \frac{1}{(ad)} - \frac{1}{(ab)}$$

$$\Leftrightarrow \frac{1}{(ac)}, \frac{1}{(ab)}, \frac{1}{(ad)} \text{ form arithmetic progression}$$

$\Leftrightarrow (ac), (ab), (ad)$ form harmonic progression.

18. The restriction of f to pc is an involution with fixed points c and a; use Theorem 3.4.10. Thus $f(p)$ is the harmonic conjugate of p relative to c and a.

Exercises 4.1

1. Let p and p' be two pencils, and let U be a line not on p or p'. The function $f: p \to p'$ defined by $f(L) = p'(U \cap L)$ (the join of p' and $U \cap L$) is a *perspectivity* from p to p' with *axis* U.

3. If $f: p \sim p'$ is a projectivity between distinct pencils in a Pappian plane and $f(pp') = pp'$, then f is a perspectivity.

5. Each singular case can be checked directly; for the nonsingular case, see the proof of Theorem 4.2.2.

7. If no three are collinear, use Theorem 1. If some three are collinear, there is a unique conic by the dual of the fundamental theorem; in each case, pq corresponds to itself, so the projectivity is a perspectivity and the conic is singular.

9. Use Theorems 1 and 2.

Exercises 4.2

2. Let p and some other point q be generating bases. Then $f^{-1}(pq)$ is unique, so there is a unique tangent at p.

4. If a, b, c, d, e are the points, with b the point at which the tangent is to be constructed, find the image of ab under the projectivity $a(c, d, e) \sim b(c, d, e)$.

7. Let L be any line not on p; then $L \barwedge p \overset{f}{\sim} p \barwedge L$ is a projectivity from L to itself. Hence there is a point s which is fixed, so ps is fixed under f.

8. Use Exercise 7.

10. By Exercise 9, every line meets every singular point conic. This, together with Theorem 3, gives the result.

12. Use Exercise 2.1.13.

Exercises 4.3

1. If the conic has equation $ax_1^2 + bx_1x_2 + cx_2^2 + dx_1x_3 + ex_2x_3 + fx_3^2 = 0$, and $[1, i, 0]$ is on the curve, then $a + bi - c = 0$, so $a = c$ and $b = 0$. Hence the conic is a circle.

3.

$$X\begin{pmatrix} 2 & -1 & 1 \\ -1 & 3 & 0 \\ 1 & 0 & 1 \end{pmatrix}X^{\mathrm{T}} = 0.$$

4. $\langle 1, -1, 0\rangle, \langle 5, 4, 5\rangle, \langle 2, -5, 2\rangle$.

6. **(a)** $[-1, 1, 1]$. **(b)** $[0, 3, -4]$. **(c)** Each point on line $\langle 1, 0, 1\rangle$.

8. If $uA = (0, 0, 0)$ and $vA = (0, 0, 0)$, then $(\lambda u + \mu v)A = (0, 0, 0)$ for all λ, μ.

10. **(a)** $\langle 0, 1, 0\rangle$. **(b)** $\langle 0, 0, 1\rangle, \langle 0, 1, 0\rangle$ **(c)** $\langle -4-\sqrt{2}, 4+2\sqrt{2}, -\sqrt{2}\rangle$.
(d) $\langle 0, 0, 1\rangle$.

12. $ax_1^2 + bx_2^2 + cx_3^2 = 0$.

14. $[4, 2, -5]$.

16. Yes.

18. q is on the polar of $p \Rightarrow pAq^{\mathrm{T}} = 0 \Rightarrow qAp^{\mathrm{T}} = 0 \Rightarrow p$ is on the polar of q.

Exercises 4.4

1. Any two secants of the circle through the given point create a four-point on the circle, one of whose diagonal points is the given point. The line on the other two diagonal points is the polar.

3. There are two or none, according as the involution induced on the ideal line has two or no fixed points.

Exercises 4.5

1. 60. See Edge [9] for a thorough description.

3. If a hexagram is circumscribed about a circle, then the diagonals of the hexagram are concurrent.

6. Let a be the point on L, let T be the other line, let the point on T be both b' and c, and let a' and b be the other two points. Let $c'' = ab' \cap a'b$, $b'' = L \cap a'c$, $a'' = T \cap b''c''$, and $c' = ba'' \cap L$. Then c' is the second intersection of L with the conic.

8. Let S and T be the tangent lines, let $a = b'$ be on S, and let $c = a'$ be on T. Let b be the third point, and suppose L is a second line on a (other cases are similar). Let $c'' = S \cap a'b, b'' = L \cap T, a'' = b'c \cap b''c''$, and $c' = ba'' \cap L$. Then c' is the second intersection of L with the conic.

10. If a, c are three points on a nonsingular conic and S, T, U are lines tangent to the conic at a, b, and c, respectively, then points

$$a'' = ac \cap T, \qquad b'' = ab \cap U, \qquad c'' = bc \cap S$$

are collinear.

12. The Pascal line of points a, b, c and their tangents is the axis of the couple.

14. Let L and M be the asymptotes. Let l be the point on L such that $pl \parallel M$, and let m be the point on M such that $pm \parallel L$. Then the tangent line at p is parallel to lm.

16. Let L and M be lines joining p to two of the five points, and construct the second intersections of L and M with the conic. This gives us a four-point on the conic, of which p is one diagonal point; use Theorem 4.4.1

Exercises 4.6

2. Let triangle pqr have polar triangle PQR. Let $pq \cap R = r_1$, $pr \cap Q = q_1$, $qr \cap P = p_1$. Also let $p' = pq \cap Q$, $r' = qr \cap Q$. Then the projectivity $(r_1pp'q) \sim (pr_1qp') \sim (PR_1QP') \barwedge (p_2rr'q)$ is a perspectivity, since q is self-corresponding. Hence p_1r_1 is on $pr \cap p'r' = pr \cap Q = q_1$, and an axial couple is formed.

4. Let $ab \cap P = x$, $cp \cap P = x'$. Since x is on ab, X is on $A \cap B = c$. Since x is on P, X is on p. Thus $X = cp$, and $X \cap P = x'$. Hence $x \to x'$ under the involution induced on P. The same is true of the other pairs in the quadrangular set. Hence by Exercise 3.4.11, the result follows.

6. In the notation of Exercise 3.1.10, if a_0, a_1, a_2, a_3 are self-conjugate, then the polarity induces an involution on a_0a_1 in which a_0 and a_1 are fixed points. Hence the image of d_1 is $h_1h_1' = d_2d_3$, and similarly for d_2 and d_3.

8. If and only if M^2 is a scalar matrix.

Exercises 5.1

2. **(a)** $i(\pi/2 + 2n\pi)$. **(b)** $-i(\pi/2 + 2n\pi)$. **(c)** $\ln 2 + i(\pi + 2n\pi)$.

3. **(a)** $\cosh iz = \frac{1}{2}(e^{iz} + e^{-iz}) = \frac{1}{2}[\cos z + i \sin z + \cos(-z) + i \sin(-z)]$
$= \frac{1}{2}(2 \cos z) = \cos z$.

8. The polar of p must meet L. Dually, some line on p is conjugate to L.

10. If q is any point on the polar of p, then q is conjugate to p, so by Theorem 5 the distance from p to its polar is $(2n+1)k_\Gamma \pi i$.

Exercises 5.2

1. Use Theorem 5.1.3.

3. Use Exercise 5.1.9.

Exercises 5.3

5. $f_{ij} = p_i A p_j^{\mathrm{T}} = p_i p_j^{\mathrm{T}}$; $\cos d_1 = f_{23}/\sqrt{f_{22} f_{33}} = 0/\sqrt{5.6}$, so $d_1 = \pi/2$. Similarly, $d_2 = \pi/2$, $d_3 = \cos^{-1}\sqrt{\frac{2}{5}}$. This is found using part 1 of Theorem 4. By part 1 of Theorem 5, we find

$$\cos \alpha_3 = \frac{\cos d_3 - \cos d_1 \cos d_2}{\sin d_1 \sin d_2} = \sqrt{\frac{2}{5}},$$

or $\alpha_3 = \cos^{-1}\sqrt{\frac{2}{5}}$. Similarly, $\alpha_1 = \alpha_2 = \pi/2$.

Exercises 5.4

6. $d_1 = \cosh^{-1}(8/\sqrt{10})$, $d_2 = \cosh^{-1}\sqrt{2}$, $d_3 = \cosh^{-1}(3/\sqrt{5})$,
$\alpha_1 = \cos^{-1}(-1/\sqrt{2})$, $\alpha_2 = \cos^{-1}(7/3\sqrt{6})$, $\alpha_3 = \cos^{-1}(5/3\sqrt{3})$.

Exercises 6.1

2. If points a and b lie in both of two subspaces, then line ab lies in both subspaces, and hence in their intersection.

4. Use Exercise 3.

6. Use Theorem 9.

8. Let L be a line in S_n, and let H be a hyperplane with basis $\{p_0, p_1, \ldots, p_n\}$. If a is a point of L that is not in H, then $S_n = \langle H \cap \{a\} \rangle$. Thus if x is any point on L, then there is $q \in H$ such that x is on aq, by Lemma 1. Hence q is on $ax = L$ and L meets H.

Exercises 6.2

3. Points: [0, 0, 0, 1], [0, 0, 1, 0], [0, 0, 1, 1], [0, 1, 0, 0], [0, 1, 0, 1], [0, 1, 1, 0], [0, 1, 1, 1], [1, 0, 0, 0], [1, 0, 0, 1], [1, 0, 1, 0], [1, 0, 1, 1], [1, 1, 0, 0], [1, 1, 0, 1], [1, 1, 1, 0], [1, 1, 1, 1]. Planes: $x_0 = 0$, $x_1 = 0$, $x_2 = 0$, $x_3 = 0$, $x_0 + x_1 = 0$, $x_0 + x_2 = 0$,

$x_0 + x_3 = 0$, $x_1 + x_2 = 0$, $x_1 + x_3 = 0$, $x_2 + x_3 = 0$, $x_0 + x_1 + x_2 = 0$, $x_0 + x_1 + x_3 = 0$, $x_0 + x_2 + x_3 = 0$, $x_1 + x_2 + x_3 = 0$, $x_0 + x_1 + x_2 + x_3 = 0$.

4. The lines can be described by using either pairs of points or pairs of planes. There are $\binom{15}{2} = 105$ of them.

5. $q(q^n - 1)(q^{n+1} - 1)/2(q-1)^2$.

7. An S_k is the intersection of $n - k$ hyperplanes. Thus, thinking of S_n as $S_n(D)$ for some D, S_k is defined by $n - k$ equations and S_l is defined by $n - l$ equations. Hence $S_k \cap S_l$ is defined by $n - k + n - l = n - (k + l - n)$ equations, so it is an S_{k+l-n}.

8. **(a)** Three planes and their lines of intersection. **(b)** A tetrahedron.

Exercises 6.3

5. See Zariski and Samuel [20, p. 3].

8. **(a)** (1, 0, 1, 0). **(b)** (0, 0, 0, 1). **(c)** (1, 0, 0, 0). **(d)** (0, 0, 1, 1).

13. Let $h : S_n(D) \to S_n(D)$ and $h' : S_n(D') \to S_n(D')$ be projective collineations such that $h'f = gh$. Theorem 2.7.6 generalizes to $S_n(D)$, so that h and h' are matrix-induced. Let $h = f_M$ and $h' = f_N$ with $M = (m_{ij})$ and $N = (n_{ij})$. Now g is basic, so $g(m_{i0}, m_{i1}, \ldots, m_{in}) = (n_{i0}, n_{i1}, \ldots, n_{in})$ for $i = 0, 1, \ldots, n$. Now we may assume that all the m_{ij} are in A_{ϕ_g} but not all in J_{ϕ_g}. Then the same is true of some row of M, say the first. If $a \in J_{\phi_g}$, then

$$\begin{aligned}
h'f(1, a, 0, \ldots, 0) &= gh\,(1, a, 0, \ldots, 0) \\
&= g((1, a, 0, \ldots, 0)M) \\
&= g(m_{00} + am_{10}, m_{01} + am_{11}, \ldots, m_{0n} + am_{1n}) \\
&= (\phi_g(m_{00} + am_{10}), \ldots, \phi_g(m_{0n} + am_{1n})) \\
&= (\phi_g(m_{00}), \ldots, \phi_g(m_{0n})) = g(m_{00}, \ldots, m_{0n}) \\
&= (n_{00}, \ldots, n_{0n}).
\end{aligned}$$

Hence $f(1, a, 0, \ldots, 0) = h'^{-1}(n_{00}, \ldots, n_{0n}) = (n_{00}, \ldots, n_{0n})N^{-1} = (1, 0, \ldots, 0)$. But $f(1, a, 0, \ldots, 0) = (\phi_f(1), \phi_f(a), \ldots, \phi_f(0))$, so $\phi_f(a) = 0$ and $a \in J_{\phi_f}$. Hence $J_{\phi_g} \subseteq J_{\phi_f}$. Similarly, $J_{\phi_f} \subseteq J_{\phi_g}$. Thus $\phi_f = \phi_g$, and $f = f_{\phi_f} = f_{\phi_g} = g$.

References

1. Agnew, J. and Knapp, R. C. *Linear Algebra with Applications*. Monterey: Brooks/Cole, 1978.
2. André, J. Über Homomorphismen projektiver Ebenen, *Abb. Math. Sem. Univ. Hamburg* **34**(1969/70), 98–114.
3. Ayers, F. *Theory and Problems of Projective Geometry*. Schaum's Outline Series. New York: McGraw-Hill, 1967.
4. Bruck, R. H., and Ryser, H. J. The nonexistence of certain finite projective planes, *Canadian J. Math.* **1**(1948), 88–93.
5. Coolidge, J. L. *A History of Geometrical Methods.* Oxford: Clarendon Press, 1940.
6. Court, N. A. *College Geometry.* College Outline Series, New York: Barnes and Noble, 1952.
7. Dembowski, P. *Finite Geometries.* New York: Springer-Verlag, 1968.
8. Dress, A. Metrische Ebenen und projektive Homomorphismen, *Math. Z.* **85**(1964), 116–140.
9. Edge, W. L. A footnote on the mystic hexagram, *Math. Proc. Camb. Phil. Soc.* **77**(1975), 29–42.
10. Garner, L. E. Fields and projective planes: a category equivalence, *Rocky Mtn. J. Math.* **2**(1972), 605–610.
11. Herstein, I. N. *Topics in Algebra.* 2nd ed. Lexington, Mass.: Xerox College Publishing, 1975.
12. Hughes, D. R., and Piper, F. C. *Projective Planes.* New York: Springer-Verlag, 1973.
13. Klingenberg, W. Projektive Geometrien mit Homomorphismus, *Math. Annalen* **132**(1956), 180–200.
14. Radó, F. Non-injective collineations on some sets in Desarguesian projective planes and extension of non-commutative valuations, *Aeq. Math.* **4**(1970), 307–321.
15. Springer, C. E. *Geometry and Analysis of Projective Spaces.* San Francisco: W. H. Freeman, 1972.
16. Stevenson, F. W. *Projective Planes.* San Francisco: W. H. Freeman, 1972.
17. Veblen, O., and Bussey, W. H. Finite projective geometries, *Trans. Amer. Math. Soc.* **7**(1906), 241–259.

18. Veblen, O., and Young, J. W. *Projective Geometry.* Volume I. Boston: Ginn, 1938. Volume II. New York: Blaisdell, 1946.
19. Verdina, J. *Projective Geometry and Point Transformations.* Boston: Allyn and Bacon, 1971.
20. Zariski, O., and Samuel, P. *Commutative Algebra.* Volume II. Princeton: D. Van Nostrand, 1960.

Index